QUANTITATIVE ASSESSMENT OF POLLUTION IN THE SANITATION, WASTE MANAGEMENT, AND ENERGY SECTORS

FEBRUARY 2026

ASIAN DEVELOPMENT BANK

ADB

6 ADB Avenue, Mandaluyong City, 1550 Metro Manila, Philippines
Tel +63 2 8632 4444; Fax +63 2 8636 2444
www.adb.org

ISBN 978-92-9277-665-7 (print); 978-92-9277-666-4 (PDF); 978-92-9277-667-1 (e-book)
Publication Stock No. TCS260030
DOI: http://dx.doi.org/10.22617/TCS260030

Notes:
In this publication, "$" refers to United States dollars.
ADB recognizes "China" as the People's Republic of China and "Turkey" as Türkiye.

Cover photos by ADB.

CONTENTS

TABLES, FIGURES, AND BOXES

Tables

Figures

Boxes

ACKNOWLEDGMENTS

This study was initiated and authored by Xin Ren, senior environmental specialist at the Asian Development Bank (ADB), and was funded under the regional technical assistance (TA) project, Improving the Readiness and Implementation of Safeguards in ADB-Financed Projects (TA 10491). Xin Ren has worked closely on energy and rural development projects in Southeast Asian countries, focusing on environmental management. Before joining ADB, she worked at the World Bank on diverse projects, including urban sanitation, waste management, energy, and transport sectors. Apart from her 18 years in multilateral development banks, she worked for years in the secretariat of the United Nations Framework Convention on Climate Change, the United Nations Environment Programme on waste management, and in the People's Republic of China on pollution control, hazardous waste management, and cleaner production.

The report benefited greatly from comprehensive comments and insightful suggestions from peer reviewers, Emma Frances Marsden, principal environmental specialist, and Carlito Mendoza Rufo Jr, senior environmental specialist, both from the South Asia Division of the Office of Safeguards, ADB; Jitendra K. Singh, water supply and sanitation specialist, and Sing Cho, senior urban development specialist (waste management focal), from Sectors Department 2, ADB. Publication and dissemination of this report would not be possible without the continuous support of Antoine Morel, coordinator for environmental work and TA 10491; Jocelyn Erlinda S. Munsayac, director of the Southeast Asia Division of the Office of Safeguards; and last but not the least, Nianshan Zhang, head of the Office of Safeguards. The author remains responsible for all residual shortcomings.

ABBREVIATIONS

ADB	Asian Development Bank
BOD_5	5-day biochemical oxygen demand
CH_4	methane
CO_2	carbon dioxide
COD	chemical oxygen demand
Dioxin (or PCDD/F)	polychlorinated dibenzo-dioxins and polychlorinated dibenzo-furans
EEA	European Environment Agency
EHS	environmental, health, and safety
EIA	environmental impact assessment
ESP	electrostatic precipitator
FGD	flue gas desulfurization
GHG	greenhouse gas
IEA	International Energy Agency
IFI	international financial institution
IPCC	Intergovernmental Panel on Climate Change
ISO	International Organization for Standardization
MEE	Ministry of Ecology and Environment (People's Republic of China)
MSW	municipal solid waste
NH_3-N	ammonia nitrogen
NO_x	nitrogen dioxides
OuE	European odor unit
PRC	People's Republic of China
PM	particulate matter
SO_2	sulfur dioxide
TP	total phosphorus
TSP	total suspended particulate
TSS	total suspended solids
US EPA	Environmental Protection Agency (United States)
WHO	World Health Organization
WtE	waste to energy

WEIGHTS AND MEASURES

d	day
GJ	gigajoule
h	hour
kg	kilogram
m^3	cubic meter
mg	milligram
OuE	European odor unit
t	ton
wt%	weight percent

EXECUTIVE SUMMARY

The energy and urban sectors constitute the bulk of infrastructure projects by multilateral development banks and are pollution intensive. Quantifying pollution is indispensable for determining compliance and the need for pollution control, choice of pollution control technologies, as well as for predicting the impacts on environmental quality. All are essential for informing the decision-making of a project or investment and improving the design.

However, quantifying pollution at the source or discharge point has been overshadowed by modeling its dispersion through the environment, particularly in the environmental impact assessments (EIA) of projects. As pollution at source is the input to air or water quality modeling, it is paramount to get it right, hence the objective of the study.

Major methods widely accepted in both developed and developing countries are discussed, i.e., analogy, emission factors, mass balance formula based on the law of mass conservation, and actual measurement. The first three are indirect estimations using input to a system (i.e., a project or facility) to infer pollution as part of the outputs. The actual measurement is the ultimate proof or disproof for the validity of other methods, including computerized modeling.

All these methods originate from actual measurements: analogies are based on actual pollution measured in similar projects in similar circumstances to infer the pollution of specific projects. Emission factors are an extension of analogy, drawing on many similar projects or facilities over the years and summarizing them into a unit discharge of pollution. Formulas (and models) capture the numeric relations extracted from actual measurement and academic research. As such, the operation and maintenance practices are already embedded in these methods, rendering them not necessarily inferior to computerized modeling. However, they are much cheaper and less demanding in terms of human capacity, making them more suitable at the project level and in developing countries.

Point source pollution has been better understood, with quantification being more mature and widely practiced; hence the focus of this study, although nonpoint sources are touched upon. The study covers air and water pollution as well as solid waste in urban sanitation (mainly wastewater treatment), municipal waste management (landfill, biotreatment, and incineration), and the thermal energy sector.

Quantitative assessment or prediction is performed against a baseline or standard. Chapter 2 discusses types of baselines, methods for baseline surveys and sample tests, and typical pitfalls in EIAs. It goes on to provide standard methods used in sanitation design to estimate water consumption, wastewater generation, typical characteristics of sewage, and the rules of thumb for estimating sewage treatment sludge.

For municipal solid waste (MSW) landfills, Chapter 3 presents typical characteristics of leachate in developing countries and rules of thumb for estimating the leachate amount based on a literature survey.

The main gas emissions from landfills include methane, a major greenhouse gas (GHG). To estimate landfill methane emission, the widely used formula recommended by the Intergovernmental Panel on Climate Change for the GHG inventory is also introduced. An alternative approach from a study of landfills in the People's Republic of China (PRC) is cited, as it used country-specific data to draw methane (CH_4) emission factors. Given that the country spans temperate and subtropical zones, these emission factors enable quicker and easier estimation at the project level in similar countries or regions if local data is hard to obtain.

Chapter 4 covers the biotreatment of waste (i.e., digestion and composting) in terms of estimating air emissions and water pollution, mainly by emission factors. Compiled by the European Environment Agency (EEA) to estimate project and facility air emissions, these emission factors cover typical air pollutants such as sulfur dioxide (SO_2) and nitrogen oxides (NOx) for subsectors including waste digestion, landfill, composting, and incineration. The study also presents typical characteristics and concentration ranges of digestion and composting wastewater, as well as their amount.

For the energy sector, only the pollution-intensive segment (thermal energy production) was studied. Estimating main pollutants in flue gas emissions involves several major methods, including the mass balance formula for estimating SO_2, particulate matter, and heavy metals; analogy for estimating NOx, and emission factors for pollutants from all types of fuel. Waste from the combustion process can be estimated using the same formulas in reverse, as illustrated and explained in Chapter 6. Quantifying major streams of wastewater proved more challenging, as the literature is difficult to decipher. Thus, the results listed from various sources should be treated with caution.

Incineration is essentially a combustion process, but using MSW as fuel and it can employ the same methods to estimate its pollution. However, the heterogeneous fuel leads to some unique features in their pollution, e.g., wastewater from incineration is more complex and difficult to treat (Chapter 5). Its flue gas emissions—while lower in SO_2 and particulate matter—contain more harmful pollutants such as dioxins. Incineration waste estimations—including fly ash and bottom slag—are also discussed.

The study concludes with an interesting phenomenon observed in some EIAs, which model the numeric limits in emission standards, not the estimated or actual emissions themselves, to predict the impacts or air quality on receptors or in the environment. This naturally led to overestimation in some and underestimation in others. It highlights the confusion surrounding what is truly modeled by air or water quality modeling and what constitutes the appropriate input for such modeling, as well as the importance of obtaining that input as accurately as possible, which is central to this study.

CHAPTER 1

Introduction

Quantifying pollution—especially point source pollution—is required in environmental impact assessments (EIAs) in many countries, including developing countries, for good reasons. The primary purpose of an EIA is to aid decision-making, i.e., approve a project or investment, mainly by illustrating whether the project proposed can meet environmental standards. Since environmental standards for discharge and ambient are both quantitative—mostly concentration-based, with some also load-based—EIAs have been tasked to estimate pollution quantitatively.

Secondly, more countries require proposed development or projects to comply with discharge standards, which are applied at the point of their discharge (e.g., stack or outfall), but also to demonstrate they will not cause damage to air or water quality. Therefore, dispersion modeling has been widely employed in EIAs to simulate and predict the pollution level (usually in concentration) after dispersion through air or water. At the receiving end at various distance from sources or discharge points, ambient air or water quality standards apply (Figure 1).

Figure 1 depicts that the input or starting point for modeling is emissions at source (stack). Obtaining accurate input is crucial for air or water quality modeling, as inaccurate input leads to incorrect outputs from any modeling. However, the quantification of emissions or effluent at source has received less attention than ambient quality modeling and is even overlooked in some EIAs. Yet it can be and has been done, sometimes routinely, by EIA preparers. In contrast, EIA preparers and reviewers generally have much less understanding of computerized modeling and have to trust modelers.

Objectives and Scope

This study aims to improve the pollution assessment in projects, more specifically, to strengthen pollution quantification in EIAs. Most environmental professionals conducting or reviewing EIAs are not experts in the sector concerned. Yet pollution is determined predominantly by technical processes and their scale. By providing emission factors, formulas, norms and rules of thumb, this study aims to equip environmental professionals with tools to estimate pollution and to judge the validity of the pollution-related results in feasibility studies and project design even if they are not experts in that sector.

This study covers methods commonly used and widely accepted in both developed and developing countries, not those that are still largely explored among academia. Consequently, it focuses on quantifying point source pollution where methods are well established with more consensus on good practice. Non-point source pollution or fugitive emissions will be touched upon to the extent possible. The study is concentrated on quantifying pollution at the source or discharge point, not after dispersing through environmental media which is better captured by computerized models.

Figure 1: Emission at Sources as Starting Point/Input for Dispersion Modeling

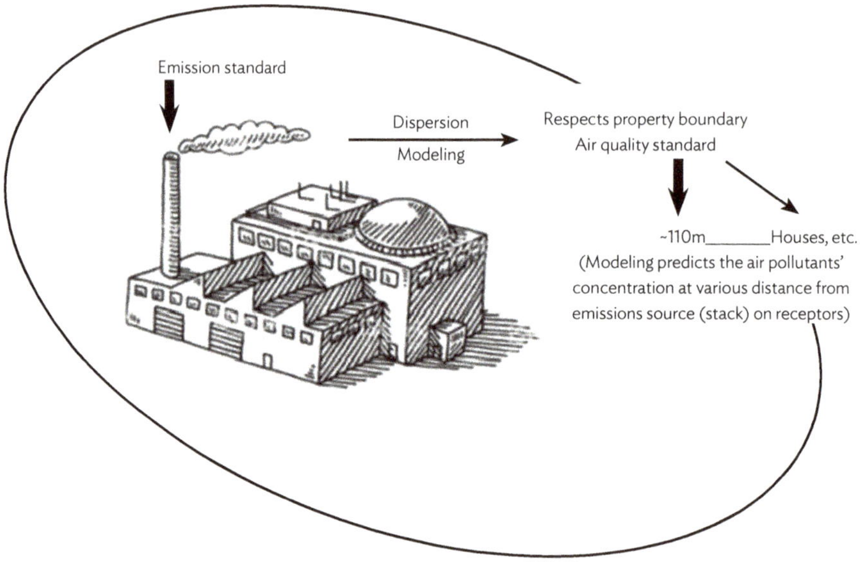

Source: Adapted from a booklet on Popular Science Modeling by the Government of Canada, Ministry of Environment (2022).

Another reason is that pollution compliance in all countries is enforced against discharge standards first and foremost at the source or discharge point, not after dispersion at the receiving end, where ambient standards will apply. This approach allows for attributing pollution to polluters, laying the basis for law enforcement, whereas results monitored at receptors can hardly be attributable to individual polluters. The differences between these two main types of environmental standards are illustrated in Figure 1 and detailed in the 2023 Asian Development Bank (ADB) study in this series.[1]

Major Approaches to Quantifying Pollution at Source

The approaches for quantitative pollution estimation commonly used in EIAs, feasibility studies, and in environmental management mainly include (Environment Australia 1999; Government of the People's Republic of China [PRC], Ministry of Ecology and Environment [MEE] 2018):

(i) analogy,
(ii) pollution or emission factors,
(iii) mass balance formula, and
(iv) actual measurement.

---

1 ADB. 2023. *Pollution Control Technologies for Small-Scale Operations.*

Analogy means using the pollution data of similar facilities to reasonably infer the likely pollution of the proposed project, such as pollution characteristics, concentration of main pollutants, and amount and/or scale. Similar facilities refer to those in the same subsector with similar scale and as similar as possible in major technology used, surroundings, material and fuel used, pollution control, and operation practice and management.

Provided that similar cases are accessible, analogy usually demands fewer resources or specialized skills. It is not necessarily less precise than more complex methods, such as modeling, as the example illustrated in the last chapter of the 2024 ADB study of this series on pollution.[2] By extrapolation and analogy, one can also roughly predict pollution before and after treatment. This is particularly useful for projects and pollutants that find it difficult to employ other methods.

When similar cases are hard to find or beyond the reasonable reach of a project, the EIA preparers can resort to **pollution and emission factors** in literature. It is also called the pollution coefficient in some countries or unit discharge, as in the 2024 ADB study (footnote 2). **Emission factor** is defined as a representative value that attempts to relate the quantity of a pollutant released with an activity associated with the release of that pollutant (United States [US] Environmental Protection Agency [EPA] 2025). In most cases, these factors are simply averages of all available data of acceptable quality, "compiled from source test data, material balance studies, and engineering estimates." Therefore, they are basically an extension of analogy at scale of many similar projects and facilities over years. As such, operation and maintenance practice is embedded in both analogy and emission factors, rendering them not necessarily inferior to modeling. But they are much less costly and easier to use, suitable for developing countries.

For better quality and consistency in emission factors, environmental authorities in some countries (e.g., the US) or regions (e.g., the European Union) conducted surveys and compiled country- or region-specific emission factors by sector. Though a great help for EIA work, the emission factors are developed mainly for environmental management and pollution census at the sector, national, or regional level. What gets measured gets managed, or better managed.

Mass or material balance derived formulas are based on the law of conservation of mass, a fundamental principle in physics and chemistry. It states that the total amount of mass in a closed system is constant over time, i.e., Total mass in = Total mass out + Total mass in the system. It is manifested in water and mass balance, as quoted in the examples in Chapters 2–6. Formulas (and models) essentially capture the numeric relations extracted from actual measurement, scientific research, and experience.

All these methods use indirect estimation using inputs to a system (i.e., a project or facility) to infer pollution as one of the outputs. The **actual measurement** of pollution is the ultimate proof or disproof for the validity of other methods, including modeling. Every effort should be made to directly measure the actual pollution, especially if the project involves expanding or upgrading an existing facility. However, measurement can be difficult or impossible with the time and resource constraints of the project. Therefore,

2 ADB. 2024. *Pollution Characterization and Quantification in the Agriculture Sector.*

EIAs and feasibility studies typically employ these methods, which in effect are all derived from past measurements. Even formulas (and models) essentially capture the numeric relations extracted partly from actual measurement.

Computerized models have been used predominantly to predict pollution at the receiving end after dispersion, which is beyond the scope of this study. Even when more accuracy can be achieved by modeling (e.g., landfill leachate estimates in Chapter 3), value for money remains a balancing act and consideration, especially at the project level and in developing countries constrained in resources, time, and capacity. Even developed countries have prioritized simpler and more straightforward methods, such as emission factors (European Environment Agency [EEA] 2019).

Methodology and Sector Coverage of the Study

This study covers major methods in pollution quantification and their application in urban sanitation (wastewater treatment and sludge treatment), waste management (landfill, biotreatment, and incineration), and the energy sector.

Energy projects usually comprise mainly thermal power plants or combined generation of heat and power, renewable energy (hydropower, wind farms, solar panels), transmission lines and substations, and energy efficiency in various industries. The last concerns pollution from industries and deserves a separate endeavor. Biological impacts dominate renewable energy and transmission line projects. This study focuses on the pollution-intensive segment of the energy sector: thermal energy production.

The **methodology** of this study is primarily a literature survey and research, supplemented by decades of experience of the author and interviews with experts. The research entailed cross-checking, calculating, and analyzing information in literature to understand, for example, which formulas are sensible yet not overly complex, whether data from different literature make sense, and the reasons for the variation in literature for the same pollution. It is particularly difficult to decipher data on streams of wastewater, as sources often use the same name for different streams or their combination; some data are often unclear, even incomprehensible, or do not add up.

The new safeguard policies of ADB, the Environmental and Social Framework, and its Environmental and Social Standard 3 on pollution and resource efficiency, require quantifying pollution to the extent possible. The 2024 ADB study on pollution provides methodology and reference values for quantifying pollution in agriculture (footnote 2). This study expands on similar efforts in the urban sanitation, waste management, and energy sectors. Together, this series aims to provide comprehensive coverage of major sectors and contribute to pollution control and environmental sustainability.

CHAPTER 2

Quantifying Pollution from Urban Sanitation

According to the World Health Organization (WHO), sanitation refers to conditions that promote health through proper collection, treatment, and disposal of human and animal waste, as well as sewage. By this definition, the technical design and feasibility study of sanitation projects must estimate the amount and strength of these pollutants. Consequently, EIAs for these types of projects do not need to repeat such work but serve as a second opinion to verify and fill the gaps.

The 2024 ADB study on pollution in this series includes characterizing and quantifying animal waste (footnote 2). This study focuses on urban sanitation, specifically sewage and wastewater. Sewage, also referred to as domestic or general wastewater in some countries, refers to wastewater from human living and working areas.

Water Pollution Versus Water Quality

As with air pollution, countries have required major development or projects to comply with effluent or wastewater standards enforced at the discharge point (e.g., wastewater outfall) and to show their potential damage to water quality. Therefore, diffusion modeling has been employed in EIAs to simulate and predict the pollution level (usually in concentration) after degradation, hydrolysis, dilution, and diffusion through water at the receiving end in various distances where water quality standards will apply (Figure 2).

Figure 2 illustrates that the starting point for water quality modeling is water pollution at the point of discharge. Likewise, correctly estimating wastewater discharge is crucial for modeling water quality downstream, the central task of this chapter on sanitation. Details and insights on water quality standards versus discharge standards can be found in the 2023 ADB study of this series on pollution (footnote 1).

In many urban and peri-urban areas, a significant portion of wastewater is managed through septic tanks and other non-sewered sanitation systems. While essential for communities lacking a centralized sewer system, these systems can be major sources of water pollution if not properly designed, operated, and maintained. Effluent from septic tanks contains high concentrations of organic matter, nutrients, and pathogens, which can seep into groundwater or discharge into surface water bodies. Incorporating these sources into pollution estimates is crucial for comprehensive water quality management and for ensuring that sanitation interventions effectively protect public health and improve the environment. Such an approach is embodied in the ADB initiative, Citywide Inclusive Sanitation, particularly in rapidly growing urban areas.

Figure 2: Diagram for Transport and Fate of Contaminants in Surface Water

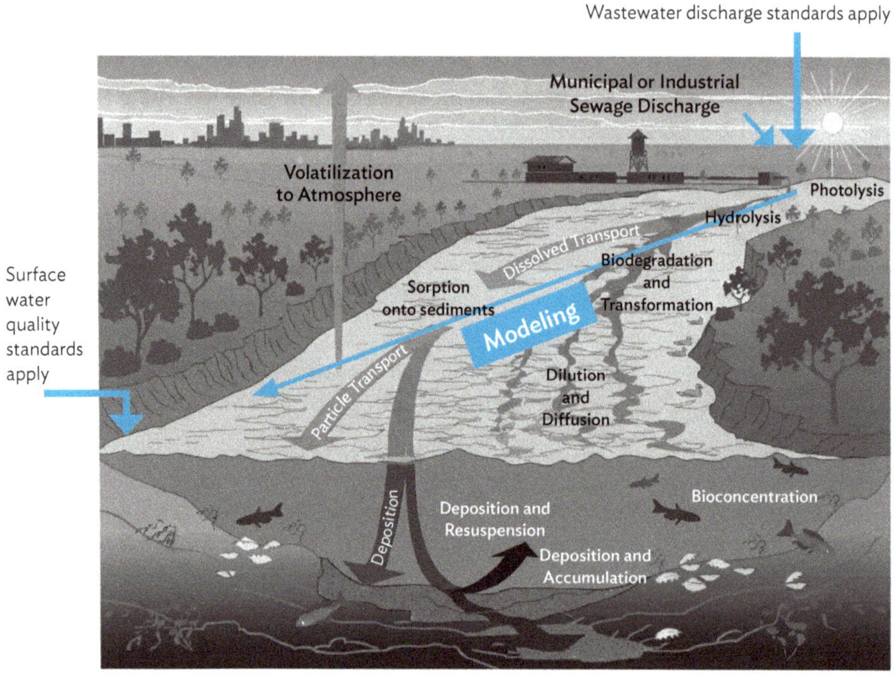

Source: Adapted by the author from Urban Watersheds: Geology, Contamination, Environmental Regulations, and Sustainability. Nd. Transport and Fate of Contaminants in Surface Water.

On a broader scale, point sources and semi-point sources of water pollution, such as non-sewered sanitation, contribute to less than half of the water pollution in many countries. The major sources turn out to be non-point or area source pollution, mostly surface runoff from farmland and agriculture. For example, experts in some developing countries estimate that non-point sources pollution dominated by agriculture accounts for about two-thirds of total pollution in surface water, with the remaining mainly from point sources. This explained why the water quality of some rivers has hardly improved despite heavy investment in wastewater treatment plants in their catchment areas. Yet quantifying these non-point source pollution has proved much more challenging than quantifying point sources. As a result, the methodologies are less mature or practiced, especially in EIAs, and thus not the focus of this study, albeit with some fugitive emissions being touched upon.

Baseline for Pollution Assessment

It is not the EIA's purpose to conduct exhaustive studies on all the impacts of projects. Scoping is thus conducted to ensure that detailed predictions are only carried out for important issues, in addition to determining spatial and temporal coverage. Detailed predictions usually intend to quantify as much as possible against the baseline situation and/or applicable standards.

There are mainly two types of baselines. One is the conditions before the project: the status and quality of air, water, soil, flora, and fauna; a snapshot at the time of project EIA preparation. Another type is a dynamic baseline (also termed reference scenario): what would happen without the project, which is not static but evolves over the same period of the project cycle. Such a reference scenario is usually obtained by extrapolating past trends into a time in the future, e.g., the would-be population, water consumption, or waste generation without the project. For most project feasibility studies and EIAs, the baseline is a combination of both and considers other approved development plans and known investments.

As part of the baseline survey, EIA preparers usually need to obtain pollution data (volume and concentration) for the existing facility or—for greenfield projects of similar facilities—production and treatment processes of similar scale in similar circumstances to infer the potential pollution of the project, essentially employing analogy. Such data are obtained from surveys and, if unavailable or insufficient, from tests conducted at similar facilities during baseline surveys of the project.

To ensure accurate quantification of pollution in EIAs, the selection and application of appropriate sampling methods are essential. Two widely used approaches are grab sampling and composite sampling. Grab sampling involves collecting a single sample at a specific time and location, providing a snapshot of conditions at that spot and moment. This method is needed for assessing air and water quality as well as localized pollution events (e.g., peak contaminant concentrations or operational upsets). In contrast, composite sampling combines multiple samples taken over a defined period or across different locations to create a representative average. This approach is particularly useful for evaluating long-term trends, spatial variability in pollution levels (such as daily average concentrations in wastewater), or less homogeneous types of pollution, such as solid waste in a city and sediment across a site.

The choice between grab and composite sampling should be guided by the project's objectives, regulatory requirements, and the nature of the pollutants being assessed. For sanitation, septage, and waste management, composite sampling can help capture spatial and temporal variability, while grab sampling may be preferred for rapid screening or compliance checks. Proper documentation of sampling protocols, including frequency, timing, and sample handling, is critical to ensure data reliability and comparability with applicable standards.

Before deciding whether to conduct actual baseline measurements and which sample tests are needed, circumstantial evidence gathered from literature and field surveys should be studied to see if they indicate any likely contamination and spatial scope or locations to narrow down the sampling. This is particularly relevant for projects involving solid substrates like sediment or sludge, as sample preparation and testing of solid substrates are more complicated and costly than air and water samples.

The standard methods for testing solids include atomic absorption spectroscopy (for metals) and chromatography (for pesticides, organic compounds, and liquid samples). They are complicated, expensive and, not easily accessible in some developing countries either. Therefore, the parameters for baseline (also for future monitoring

Box 1: When Should Air, Water, or Sediment Be Tested and How?

Many projects—especially those involving dredging or river cleanup—have not done water or sediment tests during preparation despite the presence of industrial or urban pollution in the vicinity. Instead, they require future contractors to test heavily, e.g., sediment every 2 kilometers along that river or canal. In other projects, similar blanket sampling was conducted during EIA, costing unnecessary time and budget. Neither is a good practice nor scientifically justified, especially in developing countries facing resource constraints.

Dredged material and sludge are utilized as much as possible as backfill or soil supplements, with residues being disposed of, all of which are referred to as land application. Many countries have standards for land application of sediment or sludge with numeric limits on heavy metals and pathogens (Appendix). Only a few organic compounds or pesticides are included, as they are considered likely negligible, similar to the background level, or difficult to test.

If baseline investigation, data search, and circumstantial evidence suggest that sediment or sludge are likely contaminated beyond the level allowed by the applicable standard for land application, the project must sample and test sediment or sludge as part of the environmental impact assessment. Postponing sample tests to contractors is the last resort, because it will be too late to get the right expertise and resources needed, especially if sediment or sludge is contaminated beyond the level allowed for land application or utilization (e.g., to make brick) and thus requires special handling and disposal (e.g., in a sanitary landfill or special facility for hazardous waste). This requires professional judgment by experts where contractors are unable, and can lead to delays and complications, such as claims, cost overruns, and rebidding. In some countries, contractors conduct baseline tests before construction starts as protection in case of dispute. This does not preclude the need for a sample test during environmental impact assessment.

Source: Author.

during project implementation) need to be carefully selected, prioritizing those for which relevant standard values can be found to compare results with, and for which there is an accessible laboratory capability at a reasonable cost. Many developing countries set forth sampling protocols, testing methods, and laboratory calibration and accreditation, and strive to align with international practices such as those in ISO 5667 and WHO guidelines.

As with the water-related standards, some environmental professionals are confused about the soil quality standard with the land application standard for sediment and sludge, or standards on their utilization, essentially discharge standards (Appendix). They apply more stringent limits without knowing that the numeric limits in quality and discharge standards are incomparable. Those in discharge standards are always higher than those in the corresponding quality standards (making the latter appear more stringent), considering the degradation, absorption, and dilution of pollution after being released to air, water, and soil.

Estimating Water Consumption and Wastewater

Estimating water consumption and wastewater generation should begin systematically by quantifying all main sources, uses, and losses based on site-specific data and independent verification. However, it is not uncommon to find that an EIA and feasibility study accept what others confirmed or requested as the basis for design and assessment, without their opinion on whether the information or request in question makes sense in the first place. In particular, the feasibility study by name is about justifying or verifying technical feasibility. The logical first step is to estimate the resources needed by a project. The second step is to calculate the share of the project's use (e.g., water) in the total supply to prove or disprove quantitatively whether it would affect other users or ecological functions, and to what extent.

Water Consumption

There are two basic methods to predict water consumption: actual measurement and estimation by unit water use. Actual measurement is measuring water consumption over at least 12 months to cover seasonal variation. From all testing data, select the value higher than 85%–95% of the time as the volume of water used. Whether choosing 95% or 85% depends on the local water resources conditions and requirements.

When actual measurement is infeasible due to time and resource constraints, and there is no or limited existing data, water use is estimated by unit water use, timing the number of units. If local data is unavailable or insufficient, unit water use of similar countries, regions, or international sources can be used as a proxy. For domestic water use, unit consumption and hence wastewater are similar for people in the same climatic zones with comparable socioeconomic conditions.

Wastewater Volume

As with water use quantification, there are two basic approaches to quantify wastewater volume. Actual measurement is measuring over at least 12 months to cover seasonal variation. It offers greater accuracy but demands more resources and has thus been mostly adopted in developed countries or bigger projects. Even in some developed countries or regions, wastewater volume is estimated, mainly through the following two methods.

One method is to estimate indirectly using water consumption data, which is much easier to obtain and more accurate. For processes without water intake, wastewater amounts are about 80%–90% of the water used, which accounts for infiltration, leakage, evaporation, and other unaccounted water. For those with water intake, such as industrial processes, the conversion rate to wastewater varies and can be obtained from literature (footnote 2). International codes and many countries emphasize aligning treatment capacity with the receiving environment's assimilative capacity, supported by EIAs and stakeholder consultation.

The other method is to multiply the population by unit water use in design code, then convert to wastewater. Water and sanitation design in many countries still follows the conventional design code (typically at 120–150 liters [L] per person/day), which is simpler than by facility and device type. If a town with a population of 50,000 needs to build its first wastewater treatment plant, wastewater is estimated to be 6,000 cubic meters per day (m^3/d) (= 50,000 x 150/1,000 x 80%). Adding a safe margin (usually 10%–20%) and considering population growth, the plant's designed capacity should be about 10,000 m^3/d.

Some developed countries with high living standards and life habits (e.g., taking more showers) have similar or even lower unit water use in their design code, reflecting the drive to save water and increase efficiency in the water system. Design codes with higher unit water use have led to overestimates of wastewater volume and contributed to overcapacity in some countries. In many other countries, there are still insufficient wastewater treatment facilities.

Characteristics of General Wastewater

Given the problems with design codes, it is recommended to conduct sample tests during an EIA to validate the concentrations used in the design or feasibility study. The best approach is to measure wastewater samples and test them in qualified laboratories. Table 1 shows major parameters for water pollution, which are basic ones for regulatory compliance in most countries. Microorganisms are expressed in terms of total coliforms and fecal coliforms or *E. coli*, also referred to as bacterial parameters. Many developing countries have the capacity to conduct tests of these basic parameters. Additional ones are recommended (e.g., by WHO) and adopted by some developing countries, such as heavy metals and pesticides.

The per capita contribution of organic matter and types of organic matter are constant (with only a slight variation due to diet or lifestyle). Such universality in domestic wastewater explains the abundance of data on its characteristics and concentrations, and allows the application of these concentrations in sanitation design without compromising accuracy.

Table 1: Major Constituents of Typical Domestic/General Wastewater (mg/L)

Parameter	COD	BOD_5	TSS	NH_3-N	Phosphorous	Coliform, *E. coli*
Concentration	250–500	110–220	100–300	12–25	4–8	Variable

BOD_5 = 5-day biochemical oxygen demand, COD = chemical oxygen demand, *E. coli* = Escherichia coli, mg/L= milligram per liter, NH_3-N = ammonia nitrogen, TSS = total suspended solids.

Source: Author's compilation from multiple sources.

The pollution concentration of sewage collected by wastewater treatment plants can be lower than that in Table 1, mainly because of infiltration and leakage in the sewer system. The combined sewer pipelines can exacerbate the diluting effect, leading to water intake of wastewater treatment plants unable to meet the designed strength in organic load to achieve optimal operation. Therefore, during a feasibility study and EIA, sample testing should validate design assumptions and inform treatment requirements, with stakeholder and regulatory review.

Another less-used method is water balance or its simplified form, following the principle of conservation of mass to obtain the total pollution load for each key parameter. Dividing by the total volume of wastewater, their concentration can be obtained. This method can also be preferred by environmental regulators seeking to control the total load of pollution. However, this method requires data accumulated over time and can be less reliable than direct measurement.

Quantifying Sludge from Wastewater Treatment

Sludge generation from municipal wastewater treatment plants is affected by many factors, with the major ones being wastewater quality, concentration, and the type of treatment process. To better control the treatment process, sludge is measured and calculated for each main step. Countries have developed formulas for calculation, and the more refined ones usually demand more resources to obtain data.

For feasibility studies and EIA, which are at the planning stage of wastewater treatment plants, empirical estimates by rule of thumb are sufficient and normally employed. In the PRC, each 10,000 m^3 of sewage treated normally generates 6–9 tons of sludge at 80% water content (MOHURD 2017) based on the experiences of many wastewater treatment plants. Sludge with 80% water content is achievable by the sludge dehydration methods normally adopted, which are stipulated by many countries as a minimum performance standard for wastewater treatment plant sludge. One ton of such sludge is about 0.9–1 m^3 in volume.

A simple calculation shows that further drying the water content to 60% can halve the sludge volume, making it desirable for transportation, treatment, and disposal. Many sanitary landfills only accept sludge with a water content of less than 40%, which can reduce volume by two-thirds.

Air Pollution: Odor

The main concern of air pollution for sanitation is odor, which is largely associated with sludge handling. With the expansion of wastewater treatment plants in middle-income countries, the increasing amount of sludge and its disposal have become a serious issue. The sludge treatment mainly includes drying by sun, wind, or waste heat; digestion and codigestion; and co-combustion.

Countries stipulated emission standards for the main gases of odor by concentration when they are fugitive emissions, and by amount (e.g., kilograms [kg] per hour) when there are vents or chimneys, effectively a point source discharge. To check if odor emissions meet standards, portable devices can be used to sample and test in a laboratory, though the capacity to test odor gases is less commonly available than that for sulfur dioxide (SO2), particulate matter (PM) or water pollutants.

The European standard for odor concentration uses a method called dynamic olfactometry, which involves a panel of human assessors and an olfactometer to dilute odor samples with neutral air. The result is expressed in European odor units per cubic meter (OuE/m^3), indicating the dilution level at which the odor is just barely detectable by 50% of the panel. If the sample needs to be diluted 10 times with clean air so that the panel cannot detect the odor anymore, that sample has a concentration of 10 OuE/m^3.

This is a widely accepted method for odor globally, e.g., in the US and the People's Republic of China (PRC). In addition, the PRC emission standard for odor also sets limits on the main gases that constitute odor (Table 2). This is out of concern that judgments by human assessors can be easily subject to dispute. However, many localities have difficulty testing some odor gases, which become the prevailing parameters used. The standard has been undergoing revision and updating to be more consistent with the latest air quality standard, and is likely to bring it more in line with international practice, which is more practical and easier to enforce.

Table 2: Excerpts of Odor Emission Standards at Facility Boundaries

			Class II		Class III	
Pollutant	**Unit**	**Class I**	**New, rehab**	**Existing**	**New, rehab**	**Existing**
Ammonium gas	mg/m^3	1.0	1.5	2.0	4.0	5.0
Hydrogen sulfide	mg/m^3	0.03	0.06	0.10	0.32	0.60
Methanethiol	mg/m^3	0 004	0.007	0.010	0.020	0.035
Odor concentration	OuE/m^3	10	20	30	60	70

mg/m^3 = milligrams per cubic meter, OuE/m^3 = European odor units per cubic meter, rehab = rehabilitation.

Notes: Class I limits apply to protected areas and scenic areas. Class II limits apply to residential and commercial areas. Class II limits mainly apply to industrial zones and rural areas.

Source: Government of the People's Republic of China. *National Odor Emissions Standard* (GB14554-93).

While the PRC standard in Table 2 is an emissions standard for odor emitters (the PRC air quality standard does not include odor), the odor standards in some European countries found by internet search are mostly for receptors, essentially an ambient standard. For example, in Germany, the Technical Instructions on Air Quality Control 2021 sets limits for residential and mixed areas at a maximum of 10% of annual hours where odor may be perceptible. The Environment Agency of the United Kingdom recommends different limits based on the offensiveness of odor, ranging from 1.5 OuE/m^3 for the most offensive to 6 OuE/m^3 for less offensive odors at the point of receptor. Slovenia's emission standard on odor is 500 OuE/m^3 for emitters as well. These two types of standards are incomparable, though related. For a detailed explanation, see Box 1 related text and the 2023 ADB study of this series (footnote 1).

On the other hand, human senses can easily detect odors. Noses can detect odor gases at very low concentrations, which is 400 times lower than the threshold harmful to human health. Thus, efforts have been rightly prioritized to prevent and control odor.

CHAPTER 3

Quantifying Pollution of Landfills

Solid waste generally includes municipal solid waste (MSW), industrial waste, and hazardous waste, which are further divided into subcategories. Given that urban projects of international financial institutions (IFIs) often have MSW management as an integral part, this study focuses on MSW-related pollution. Most pollution occur in landfilling and incineration, with digestion and composting to a lesser degree. The private sector mostly handles waste recycling, which is relatively rare among IFI projects and thus not covered.

Like sanitation, waste management projects are about handling, treating, and utilizing solid waste and associated pollutants such as leachate. As a result, technical design and feasibility studies must quantitatively estimate the waste, its characteristics, associated wastewater and/or leachate, and air emissions. Likewise, an EIA serves as a second opinion and informs environmental regulators about the likelihood of compliance.

Water Pollution: Estimating Landfill Leachate

The major environmental threat of a sanitary landfill is the presence of leachate. In developing countries, MSW generally produces a large amount of leachate due to high moisture content. Landfill leachate is highly complex and highly concentrated, and is formed through a combination of rainwater infiltration, moisture present in waste, and waste biodegradation. The estimation of leachate maximum volume, daily average, and composition is crucial to the capacity design and technology choice of a leachate holding (if treated offsite) and treatment system.

There is a wealth of information on research and models developed for estimating leachate generation, essentially built on the principle of water balance. However, these theoretical methods and models demand data often beyond the time and resources normally afforded to a project. Therefore, empirical methods have been developed over the years and benefited from modern design and operational practices of sanitary landfills.

The standard sanitary landfill employs a nonpermeable clay liner topped with a high-density polyethene liner, making the groundwater seepage, a leachate contributor, negligible. Surface runoff is also diverted away from landfill cells by drainage ditches and other engineering measures. Daily coverage and compaction as standard operating practice further reduce rainwater input and leachate generation.

Many countries require the retention pond onsite to be able to hold leachate generated over a certain period (e.g., 3 months) in case of malfunction of the treatment system or other accidents. Most retention ponds are open air due to their size, thus direct precipitation will add leachate volume but also dilute it. In standard design and operation, the leachate collected is pumped back to the landfill and evenly distributed around the site. This can facilitate MSW degradation, reduce the treatment burden, but also accumulate and thicken leachate in the landfill.

A 2022 review of landfills in the PRC (Solid Waste Net 2022) reveals that leachate volume can be 40% or more of MSW wet weight, due to insufficient diversion of rainwater, excessive uncovered workspace, and substandard operational practices and management. Statistics from recent years show that leachate generation from many landfills exceeds the design and overloads the treatment system.

All these show the technical complexity of MSW landfills beyond the typical field of most environmental professionals who prepare or review EIAs. Fortunately, leachate estimation and treatment are central tasks of landfill technical design and execution. With the above rule of thumb and other similar or more complicated methods (e.g., Choden et al. 2022), EIA preparers and reviewers can still offer a second opinion to technical design at the feasibility study stage.

Typical Characteristics of Leachate

The composition of leachate is influenced by various factors, the most significant of which include waste characteristics, landfill age, meteorological parameters such as temperature and precipitation, landfill depth, and waste treatment processes conducted before disposal. In planning a new landfill, measuring the leachate of existing landfills in similar circumstances is preferred whenever possible, in addition to a literature survey. Only as a last resort will data from developed countries be used, since MSW in these countries can be much drier with lower organic content, both significantly affecting the amount and concentration of leachate.

Several literature reviews resulted in similar concentrations with equally wide ranges (Table 3). Particularly useful is the literature review by Lindamulla et al. (2022), who plotted leachate data from tropical regions, categorized by subregion and landfill age. The lower concentration might be owing to higher precipitation in the tropical than the temperate zone. A study in Ireland reports leachate with chemical oxygen demand (COD) of about 86,000 milligrams per liter (mg/L) and NH_3-N of about 170 mg/L (Logan et al. 2021), more resembling those in the PRC, both in the temperate-cold climate zone.

The wide range of key pollutant concentrations underscores the need for sample tests of existing similar landfills under similar conditions during baseline investigation, especially in developing countries. Nonetheless, the data in Table 3 provide a sense of the magnitude of landfill leachate pollution and the potential harm if released without proper treatment. They also equip EIA preparers and reviewers with knowledge to cross-check whether the pollution data used in the feasibility study are broadly valid and realistic, and not too optimistic to cause problems later.

Table 3: Typical Pollution Range of Landfill Leachate (mg/L)

Item	Xiang 2025	Lindamulla	Karagozoglu	MEE 2013a		MEE 2021
				Early years	Mid stage	
BOD_5	98–50,000	2,128–3,455	NA	4,000–20,000	2,000–4,000	~2,000
COD	140–157,000	6,785–7,985	973–8,863	10,000–30,000	5,000–10,000	~6,000
NH_3-N	400–6,000	200 to ~2,000	838	200–2,000	500–3,000	~1,470
TSS	900–13,333	NA	342	500–2,000	200–1,500	NA
TP	NA	NA	NA	NA	NA	~20
Chlorine	47–6,000	2,220–3,235	NA	NA	NA	NA
Lead	0.001–5	0.001 to ~1	0.013	NA	NA	~0.130
Arsenic	0.0001–0.578	0.001–0.040	NA	NA	NA	~0.070
Phenol	1.3–800	0 to ~5	NA	NA	NA	NA

BOD_5 = 5-day biochemical oxygen demand, COD = chemical oxygen demand, mg/L = milligrams per liter, NH_3-N = ammonia-nitrogen, TP = total phosphorus, TSS = total suspended solids.

Other wastewater streams primarily originate from the cleaning of facilities, equipment, and spaces, as well as staff water use, with concentrations much lower than those of leachate and comparable to those of general wastewater discussed in Chapter 2. This is also true for other waste management in Chapters 4 and 5.

Estimating Gaseous Emissions from Landfills

The primary emissions from landfills include methane (CH_4, 45%–50% by volume) and carbon dioxide (CO_2, 40%–60%). Though both are the top two greenhouse gases (GHGs), CH_4 is flammable and viewed as a pollutant, and is usually assessed in the EIA. A mixture of methane and air becomes flammable if the concentration of methane is 5% by volume or higher. The emission standards for landfills and waste digestion primarily focus on preventing explosive risks due to methane and mitigating greenhouse gas emissions.

National standards in the US for the performance of MSW landfills (US EPA 2016) require landfills with CH_4 emissions above a certain threshold (tons per year) to have a gas collection and control system. Similar requirements are outlined in the PRC national standards for pollution control of MSW landfills, specifically that methane emissions at landfill gas pipes should be less than 5% by volume.

Meeting these regulatory requirements and obligations entails CH_4 estimation in feasibility studies and EIAs, focusing on the total amount. The commonly accepted and used method of estimating landfill CH_4 is the method developed by the Intergovernmental Panel on Climate Change (IPCC) and further refined in 2019 (IPCC 2019), intended for GHG inventory. Its first-order decay model is the mainstream approach and is adopted by the US EPA and PRC guidelines as well (MEE 2013a). Therefore, GHG counting is only discussed in this chapter on quantifying landfill emissions, not in other chapters for subsectors that do not have methane as a major emission.

Methane: Counting Greenhouse Gas Emissions

Like leachate from landfill, emissions of CH_4 are highest in the first few years after landfilling, then gradually decline as the bacteria consume the degradable carbon in the waste. Laboratory and field observations suggest that the overall decomposition process can be approximated by first-order kinetics, which has been widely accepted. The IPCC has adopted a relatively simple first-order decay model to estimate CH_4 emissions by steps, each with a formula. Put together, the overall formula is expressed as:

$$E_{CH_4} = M \cdot \sum_{i=1}^{4} C \cdot f_i \cdot D_i \cdot D_F \left(e^{-(T-1).k_i} - e^{-T.k_i}\right) \cdot F \cdot 16/12 \cdot (1-R) \times (1-O) \tag{3-1}$$

where E_{CH4} is the CH_4 emitted in inventory year T (ton); M is the mass of MSW landfilled at time 0 (ton); C is the correction factor; f_i is the fraction of waste type i (kitchen waste, paper, textile, and wood, in %); D_i is the fraction of degradable organic carbon in waste type i; D_F is the fraction of degradable organic carbon; k_i is reaction constant; F is the fraction of CH_4 by volume landfill gas which is about 50% with high certainty among all parameters in the formula; R is the CH_4 recovery rate; and O is the oxidation factor reflecting CH_4 oxidation at surface of landfill.

The IPCC guidelines on waste provide data ranges for these parameters in different regions with key tables. D_F is an estimate of the fraction of carbon that is ultimately degraded and released from landfills, as some degradable organic carbon does not degrade, or degrades very slowly, under anaerobic conditions in disposal sites. The IPCC recommends a default value for D_F of 0.5. For oxidation occurring at the surface of disposed waste captured by parameter O, the IPCC recommended "0" for unmanaged sites and "0.1" for managed sites with covers such as soil or compost.

Given that these data are mainly obtained from developed countries whose MSW are quite different from those in developing countries, some developing countries investigated their own to be used in CH_4 estimation, using the tier 2–3 approaches encouraged by the IPCC. For example, PRC researchers (Cai et al. 2014) conducted a nationwide survey and found that the waste composition and their *Di* data differ notably from the default values in the IPCC guidelines (Table 4).

Table 4: Default Values Versus Data on Degradable Municipal Solid Waste (% in wet)

Parameter	Kitchen Waste	Paper	Textile	Wood	Degradable
Average share in the PRC	47.0	6.4	2.2	3.0	59
Southern PRC (sub-to-tropical)	44.0	5.8	2.0	3.2	55
IPCC default share	26.2	18.8	3.5	3.5	52
Di in the PRC	11.0	24.0	27.0	33.0	NA
Di default value of the IPCC	15.0	40.0	24.0	43.0	NA

Di = the fraction of degradable organic carbon in waste type *i*, IPCC = Intergovernmental Panel on Climate Change, NA = not applicable, PRC = People's Republic of China.

Sources: Cai et al. (2014) and IPCC (2019).

Even with all the IPCC default values, estimating the CH_4 emission of an individual landfill from year 1 for its lifespan remains a daunting task for developing countries, particularly at the project level. That is why most countries normally do not require quantifying CH_4

emissions in an EIA, but separately under GHG emission counting or climate change assessment. Some IFIs, including ADB, require an estimate of GHG emissions if a project will likely emit GHGs above a threshold per year. How does one judge if a project will exceed a threshold at the proposal stage before any assessment?

The study by Cai et al. (2014) calculated different sizes of landfills for 40 years using country-specific data whenever possible and converted the results into CH_4 emission factors (Table 5 provides an excerpt). The column "South" refers to southern PRC, which largely falls in the subtropical zone. These emission factors enable quicker and easier estimations of CH_4 at the project level and can be applied in similar countries or regions if local data is difficult to obtain.

Table 5: Landfill Methane Emission Factors in the People's Republic of China (kg CH_4/t wet MSW)

Years of Landfill	Landfill >5 Mm^3		Landfill 2–5 Mm^3		Landfill <2 Mm^3	
	South	Average	South	Average	South	Average
1	1.96	2.04	1.94	2.12	1.36	1.57
5	1.080	1.100	1.070	1.154	0.750	0.840
10	0.56	0.54	0.55	0.56	0.38	0.41

Source: Cai et al. (2014).

For example, a project proposal to build an MSW landfill of 3 million cubic meters in a subtropical region with a daily landfilling capacity of about 100 tons has emission factors in Table 5 for year 1—when CH_4 generation is highest in the landfill—is 1.94 CH_4 kg/t of MSW. The total amount of MSW landfilled in year 1 will be 36,500 t/year. Thus, the landfill is estimated to emit about 70 tons of CH_4/year (1.94x36,500 /1000). Given the global warming potential of CH_4 is 28–36 times that of CO_2 according to the IPCC, the landfill will emit about 2,000–2,500 tons of CO_2 equivalents/year in year 1, well below the prevailing threshold of IFIs for GHG counting. Therefore, this typical landfill project is exempt from the obligation of more detailed CH_4 counting using the IPCC formula (3-1). Using the less accurate tier 1 approach of the IPCC (i.e., also by emission factors) will reach a similar conclusion. ADB uses the same IPCC formula (3-1) for estimating landfill CH4 emissions (ADB 2017a).

Estimation of Odor

The air emissions from landfill and leachate retention pond contain minor amounts of hydrogen sulfide (about 0.1% by volume), ammonia (0.1%–1%), and methanethiol (1%–3%), which are the main sources of odor. As fugitive non-point source emissions, they are difficult to estimate by empirical formulas or models. Emission factors have been developed for odor from MSW landfills in the US (US EPA 2025).

CHAPTER 4

Quantifying Pollution from the Biotreatment of Waste

The biotreatment of solid waste applies to organic fractions, mainly kitchen waste, garden or green waste, agrowaste, nonrecyclable paper (without coating of persistent chemicals), and sewage treatment sludge. It includes aerobic (i.e., composting), anaerobic, and semi-anaerobic processes (i.e., digestion) with end products like fertilizer or compost, thus waste utilization. Digestion also generates biogas, a byproduct that, if collected, can be used as an energy source.

Quantifying Pollution from Anaerobic Digestion

Wastewater or Slurry of Digestion

As for any biodegradation process, the impurity of the input is detrimental to its success. Therefore, the organic fraction of MSW segregated at source still needs to be further sorted in the plant. It undergoes air flotation to remove components of varying densities, magnetic sorting to remove ferrous metals, and water swirling to crush and separate. Swirling refers to a homogenization process used in pretreatment to separate organic fractions of MSW for anaerobic digestion. This stream of water, together with water contained in the MSW organic fraction, will enter the digester and form slurry in the upper layer of the digester and be discharged so that the desirable moisture level for digestion can be reached. A pressure belt filter dewaters the remaining digestate after digestion. The liquid generated from filters is usually treated together with the slurry.

The amount of water used in pretreatment is much greater than the water contained in MSW. In the EIA for a kitchen waste biodigestion plant (Beijing 2008) with a designed capacity of 200 tons MSW/day, the water balance indicated that the water circulated in the swirling process was 800 m^3. Together with leachate and slurry, about 860 m^3/d of digestion wastewater needs treatment. With a safe margin of 10%–20%, this totals a designed capacity of about 1,000 m^3/d. The treated wastewater is recirculated with fresh water of 7–8 m^3/d added, according to the water balance. In North America, less water is added to the segregated MSW in pretreatment; about 0.5–0.7 m^3 of water per ton of MSW in wet anaerobic digestion, which is the common type in developing countries. Moreover, up to 45% of water fed into the digestor is discharged as effluent (Environment Canada 2013), much less than in the case above. This is perhaps due to better source separation of MSW and operational practices in developed countries.

Performing a mass/water balance is now a standard practice in feasibility studies for sectors such as thermal power plants, waste management espacially incineration, and industries. Water balance assumes that water input to a system during a period is equal to the output. The benefits of mass and water balance are illustrated in several cases throughout this report and in the author's experience when overseeing the environmental audit for an existing medium-sized factory.

Given that water pollution is the main concern of that factory, a water balance was conducted. It revealed that greenery is the largest water user, consuming several hundred cubic meters per day across a factory area of about 1 hectare. Such unusual results are exacerbated by the fact that the factory is in a wet region with annual precipitation of 1,000 millimeters. It was therefore suspected that water was largely used to dilute the factory's wastewater to meet discharge standards. As further proof, the water balance of the PRC digestion plant described above with a similar land area showed normal greenery and the area cleaning only consumed several cubic meters per day.

The liquid from dewatering digestate, although smaller in volume than slurry, is higher in solids, organics, ammonium nitrogen, oil and grease, or salts (Shi et al. 2022), posing a challenge for its treatment. The unique feature of kitchen waste in the PRC (high in oil and grease) provided an incentive for extracting oil and subsequently their separate collection in many cities. Some patterns have gradually emerged and been summarized in official guidelines, e.g., digestion leachate is about 80% of kitchen waste by weight (MEE 2021).

Another main source of wastewater in digestion comes from MSW water content (about 70%). Given that the belt filter can press digestate to a water content of 70%–80%, water in MSW largely remains in post-filter digestate or sludge, and is hardly released as wastewater. Further drying of dewatered digestate by process heat evaporates water to air. It is apparent that the use and efficiency of the water swirling process in the pretreatment and dewatering of digestate largely determine the scale of the digestion wastewater.

Characteristics of Digestion Wastewater

Digestion wastewater consists mainly of the upper slurry in the digester, leachate from pressure dewatering of digestate, and in-plant sorting wastewater, and is usually treated together. Digestion of MSW is similar to the passive degradation process in landfills; hence, its wastewater has characteristics similar to landfill leachate, especially that of mid- to late-stage landfills, as reported by some studies (Shi et al. 2022) and EIAs (Beijing 2008). The concentrations of digestion wastewater can be higher than those of landfill leachate, largely because landfills and retention ponds receive precipitation, which dilutes their leachate (Table 6).

Table 6: Wastewater Characteristics of Municipal Solid Waste Digestion (mg/L)

Parameter	Environment Canada 2013	MEE 2021	Beijing 2008	Shi et al. 2022[a]	ADB 2019
BOD_5	1,500–15,000	5,600	3,500	15,000–21,000	10,000–30,000
COD	NA	13,800	5,000	6,000–8,000	30,000–50,000
NH_3-N	1,000–3,000	1,600	300	2,500–3,300	1,000–2,000
TSS	50–5,000	NA	400	2,000–21,000	1,500–10,000
Oil and grease	NA	NA	200	1,000–2,800	NA
TP	NA	60	NA	470	NA
Chloride	NA	NA	NA	NA	2,000–10,000

BOD_5 = 5-day biochemical oxygen demand, COD = chemical oxygen demand, mg/L = milligrams per liter, NA = not applicable, NH_3-N = ammonia-nitrogen, TP = total phosphorus, TSS = total suspended solids.

[a] Lower numbers in this column are the designed upper limit of leachate intake of a kitchen waste digestion facility, whereas the higher numbers are the actual test data of leachate after operation of that facility.

Air Emissions from Waste Digestion

Air emissions from digestion are mainly odor from digestion and sludge dewatering processes, and biogas, which is dominated by CH_4, about 60% in volume. Odor is the main nuisance in digestion facilities, causing public complaints. Although the digester is completely enclosed so that no emissions should escape, any accidental release is relevant for GHG counting. Digestion plants that do not capture biogas need to estimate CH_4 as a major GHG emission and monitor it because of safety concerns about landfill methane, as required in the European Union and many countries.

The European Monitoring and Evaluation Program of air pollutants under the European Environment Agency (EEA) mandated European countries to inventory their major air pollutants for the implementation of the region's Convention on Long-range Transboundary Air Pollution. The approach for countrywide air pollution estimation is adopted by the worldwide GHG inventory (IPCC 2019), i.e., activity data times emission factors to estimate various air emissions.

Like the IPCC methods to estimate GHG emissions, the EEA methods and emission factors have been used to estimate air emissions at project and facility levels, since they also cover typical air pollutants such as SO_2, NOx, or PM (not just GHGs) for different subsectors, including waste digestion, landfill, and incineration. For odor gases, there is only an emission factor for ammonia for digestion and composting (EEA 2023).

$$E_{NH3} = AR_{feedstock} \times EF_{NH3\text{-}N} \times 17/14 \quad (4\text{-}1)$$

E_{NH3} is NH_3 emission (in kg NH_3 per a period); $AR_{feedstock}$ is nitrogen in feedstock consumed in the same period; $EF_{NH3\text{-}N}$ is the emission factor (in kg NH_3-N per kg N in feedstock); and 17/14 converts N to NH_3.

The default values for EF_{NH3-N} for digestion at biogas facilities, its open storage of digestate, and the nitrogen content of various feedstock are provided and have been converted to kg/t in this study for easier use in practice (Table 7).

Table 7: Emission Factors of Digestion

Pollutant	Emission Factor (kg/t Nitrogen)	Feedstock	Dry Mass in Fresh Matter (kg/t)	Nitrogen in Fresh Matter (kg/t)
NH_3 of digestion	27.5	Organic MSW	400	6.8
95 % interval of confidence	16.3–50.1	Green waste (grass)	NA	4.6
NH_3 of digestate open storage	26.6	Poultry manure Pig manure	500 250	17.5 6
95 % confidence	15.2–46.5	Cattle manure	250	5.2
		Pig slurry	60	4.8
		Cattle slurry	100	5.2

kg/t = kilograms per ton, MSW = municipal solid waste, NH_3= ammonia.

Source: European Environment Agency. 2023. Air Pollutant Emission Inventory Guidebook 2019. 5.B.2, Biological treatment of waste – anaerobic digestion at biogas facilities.

With the CH_4 emission factor for digestion (27.5 kg/t) and N content of organic MSW (6.8 kg/t), NH_3 emission can be estimated as 227 kg NH_3/ton of organic MSW (= 1 x 6.8 x 27.5 x 17/14). Based on Table 7, the digestion of poultry manure emits the highest NH_3, three times that of organic MSW.

Converting Emission Amount to Concentration

Converting the estimated emission amount or load to concentration, one needs to know the gas or air flow rate at the facility. For digestion, which is normally enclosed with a gas pipe, the gas flow can be tested (actual measurement approach) or using data from similar existing facilities (analogous approach). Therefore, CH_4 and NH_3 concentrations at discharge (essentially a point source) can be calculated and checked for compliance with standards.

For landfill and semi-open composting, which are open air, weather and topographic conditions (of landfill), design, and ventilation (e.g., for composting) will affect the air flow on their surface, thus the emissions' concentration. To estimate such non-point source or fugitive emissions, the best and easiest way is actual measurement or analogy with the monitoring data of existing similar facilities under as closely similar conditions.

Quantifying Pollution from Composting

Composting is a controlled aerobic and exothermic process to decompose organic solid waste into a humus-like compost that can be used as a soil conditioner.

Air Emissions of Composting

The main air pollution from composting is odor due to the potential anaerobic process in the otherwise aerobic process of composting. Composting itself also creates some odorous volatile products, though less than digestion in general. This is one reason composting is advocated over digestion; however, not generating methane to recover energy is also a driver to shift from composting to digestion. There is also limited information on quantifying odors from this source. The EEA 2023 guidebook on waste composting includes emission factors only for NH_3, i.e., 0.24 kg/t of organic waste as default with a range of 0.1–0.7 kg/t, and 0.66 kg/ton of garden and park waste with a range of 0.05–1 kg/t, overall about 1%–3% of NH_3 from digestion.

Composting Wastewater

Compared to digestion, composting is exothermal and generates less water. However, like digestion, in-plant storage and sorting as a pretreatment of organic waste generate wastewater with amounts and characteristics similar to digestion wastewater. In-plant sorting is needed to further remove impurities, a major hurdle to successful composting, and to crush waste into particles suitable for composting. Anaerobic conditions occur inside and at the bottom of composting piles, especially in passive-aerated composting prevailing in developing countries. This leads to leachate generated and drained from composting and curing piles, as well as rainwater runoff infiltrating.

In the PRC, a case study in a rural area (Environment Network 2023) found that composting 5 tons of wet MSW can generate 3 tons of compost and 1.8 tons of leachate (i.e., 36% of raw MSW weight), which is usually not treated and discharged directly. An earlier study on composting kitchen waste (Zhang et al. 2011) found that leachate from composting accounted for 38% of waste wet weight, similar to the rural case study. The remaining water in MSW is lost mainly through evaporation. This article and other surveys (MEE 2021) demonstrated that adding dryer agrowaste, such as stalk, can significantly improve composting efficiency by balancing the nutrients and reducing leachate generation to around 5% of waste by wet weight.

CHAPTER 5

Quantifying Pollution of Incineration

Incineration is a specific type of combustion using waste as fuel, while combustion is a broader term encompassing any process where a substance reacts with oxygen to produce heat and light. Waste incineration with energy recovery (also called waste to energy, WtE) in the form of electricity and/or heat is essentially a thermal power plant or boiler burning MSW as fuel. Incineration thus has typical wastewater and flue gas emissions of combustion (Chapter 6), but special wastewater in pretreatment, emissions of dioxins, and waste unique to its fuel.

MSW incineration remains rare in developing countries, except for a few, like the PRC, which saw a boom during 2004–2023, resulting in more than 1,000 incineration plants by 2023. Notwithstanding many issues (Ren 2021), they provided rich experience and information, including data on actual emissions, wastewater, and ratios of bottom and fly ash. Given that MSW characteristics in developing countries differ substantially from those in the developed world, this chapter is inevitably based more on literature and experiences from the PRC.

Water Pollution: Type, Scale, and Properties

Given the higher moisture levels of MSW (50%–70%) in developing countries, a common practice by many incineration plants has been to store MSW for 6–10 days in bunkers normally designed for 7 days of feedstock for semi-anaerobic fermentation as a pretreatment. This is, in effect, the same as the digestion discussed in Chapter 4, with leachate generated during waste dehydration, desirable for incineration. The amount of incineration leachate is comparable to that of digestion, excluding pre-sorting, i.e., about 20%–30% of MSW wet weight according to EIAs for incineration and guidelines for treatment and reuse of leachate from MSW incineration plants (ISO24297, 2022). They, in turn, cited actual experience of many incineration plants in various countries (Table 8).

In a World Bank project to improve MSW incineration (World Bank 2014), the existing incineration plant had a capacity of 1,000 tons per day (t/d) and a leachate treatment of 200 t/d (about 20% of MSW weight) that had been able to handle leachate. In an article on the MSW incineration plant of 600 t/d (Chen 2022), the water balance indicated 190 t/d of leachate during semi-aerobic fermentation as pretreatment. Digestion leachate in this case was about 30% of the MSW wet weight.

Table 8: Main Wastewater of Incineration

Wastewater	Weishi County 2020	Zhoukou 2020	ADB 2019[a]	World Bank 2014
Designed capacity, heat value, MSW composition	Burn: 1,200 t/d MSW	Burn: 1,200 t/d MSW; 4,526–5,200 kJ/kg; Organic 56%, dirt 18%, S: 0.1%	Burn: ~400 t/d; S: 0.45%, ash: 12.3%	Burn: 1,000 t/d (Xishan)
1. Leachate of MSW: including loading	~300 m^3/d (25% MSW weight)	~320 m^3/d (26% of MSW)	80–180 m^3/d (average 20% of MSW)	200 t/d (20% of MSW)
2. Process wastewater: (i) blowdown of cooling water (ii) demineralization (iii) boiler discharge	(i): 260 m^3/d (ii): (132 m^3/d)	(i)+(ii): 480 m^3/d (iii): 20 m^3/d	NA	NA
3. Concentrate of membrane treatment	~60 m^3/d	~80 m^3/d	NA	NA
4. General wastewater	10 m^3/d	12 m^3/d	~10 m^3/d	NA

ADB = Asian Development Bank, kJ/kg = kilojoule per kilogram, m^3/d = cubic meters per day, MSW = municipal solid waste, NA = not available , t/d = tons per day.

[a] The contractor/operator is a Chinese company known for MSW incineration.

Another main difference between incineration and combustion is the wastewater from flue gas treatment. The sulfur content in MSW typically ranges from 0.14 to 0.71 weight percent (wt%), with an average of 0.38 wt% (Cheng et al. 2023). This is generally lower than for coal and oil, though it can vary depending on the specific waste composition, with some fractions such as kitchen waste and rubber having higher sulfur content. Incineration flue gas treatment usually does not have dedicated desulfurization, just dust and NOx removal, which can co-remove sulfur and chlorine. The resulting wastewater is much smaller than that for coal and oil combustion (Table 9). Flue gas emission treatment stabilizes these gases through chemical reactions with agents injected and turns the gaseous emissions into liquid (wastewater) or solid state, such as fly ash or gypsum, not zero-discharge as often claimed.

Air Emissions Quantitative Assessment

Municipal waste incineration is known for air emissions, including nitrogen oxides (NOx), volatile organic compounds, sulfur dioxide (SO_2), and particulate matter (PM). Given the presence of halogen compounds in plastic waste, incineration has been a source of polychlorinated dibenzo-dioxins and polychlorinated dibenzo-furans, also known as dioxins, other persistent organic pollutants, and some heavy metals such as mercury. The type of incinerator, its capacity, way of operation (e.g., whether it includes heat recovery), and the degree of abatement are the main factors influencing air emissions.

Estimation by Emission Factors

The composition of MSW can be highly variable, particularly between developed and developing nations. Typically, MSW in developed countries has about 8–12 gigajoules per ton (GJ/t) heat value as received, about one-third of that for coal (25–30 GJ/t). The EEA's emission factors for incineration are based on a net calorific value of 10.5 GJ/t of MSW (Table 3-1 of EEA 2019). MSW in the US has the same heat value (US EPA 1996). This can generate about 600 kilowatt-hours of electricity (IEA 2003).

Table 9: Wastewater Properties of Incineration

Wastewater	Scale	Typical Property (mg/L)	Typical Treatment
1. Leachate of MSW: Loading, storage, pretreatment	20%–30% of MSW wet weight	Very high concentration:[a] BOD_5: ~30,000; Cr: 0.5 COD: ~60,000; Pb: 1.5 TSS: ~12,000; Hg: 0.025 NH_3-N: 2,000	Biochemical (e.g., activated sludge, MBR) followed by ultrafilter and RO
2. Process wastewater: - Cooling water - Boiler water demineralization - Cleansing boiler	1/3 ~40% of MSW wet weight	Low concentration: BOD_5: 100–200; TSS: 50–100 COD: 100–350; salt: ~3,000 NH_3-N: 10–30	Low concentration, often treated in a system separated from leachate: chemical-physical, i.e., flocculation, filtering, RO, or nanofilter
3. Wastewater from flue gas treatment	Minor	NA	
4. Concentrate of membrane tech (RO, ultrafilter)	20%–35% of liquid after RO, various filters ~3% MSW[b]	High in salts, suspended solids	Lime slurry making, fly ash stabilizing, back-spray in combustion, evaporation
5. Domestic/general: Cleaning vehicle/ area, staff use	~1% of MSW weight	See Chapter 2	Drained to sewer, septic tank, or leachate system

BOD_5 = 5-day biochemical oxygen demand, COD = chemical oxygen demand, Cr = chromium, Hg = mercury, MBR = membrane bioreactor, mg/L = milligram per liter, MSW = municipal solid waste, NA = not applicable, NH_3-N = ammonia nitrogen, Pb = lead, RO = reverse osmosis, TSS = total suspended solids.

[a] Weishi (2020), Zhoukou (2020).

[b] ADB (2019).

In contrast, MSW heat values in developing countries are much lower (e.g., averaging 4.2 GJ/t for MSW in the PRC), less than half of that in developed countries, due to higher moisture and organics like kitchen waste and a lower share of flammables like paper. This has posed a tremendous challenge for incineration operators who have added coal or oil to aid combustion with a series of consequences (Ren 2021).

Table 10 presents Europe's tier 1 method which uses default emission factors to estimate emissions from MSW incineration (EEA 2023). For the tier 1 emission factors, it is assumed that desulfurization, abatement of NOx by selective non-catalytic reduction, and particle abatement by electrostatic precipitation and/or fabric filter technologies, and activated carbon are in place. The emission factors can be assumed to be representative of modern WtE incineration plants following the implementation of the EU waste incineration directive.

Most, if not all, incineration in the PRC is WtE plants with electricity generated as their major revenue. They have all the emission abatement stipulated in the EU, from which EEA emission factors are derived. Notwithstanding the gaps in operation practice, the huge difference in heat value of MSW means incineration in developing countries like the PRC needs to burn more than double the tonnage of MSW to produce one GJ of energy than its European counterparts. Since emissions are proportional to the amount of waste incinerated, more than double the amounts of all pollutants are emitted per gigajoule of energy produced; hence, emission factors per unit energy in the PRC are more than doubling those in Europe.

Table 10: Emission Factors for Incineration

Pollutant	Old Plant, No WtE (g/t)[a]	Modern with WtE (g/t)[a]	WtE in EU (g/GJ)[b]	US EPA (g/GJ)[b]
SO_2	1,700	87	8.7	3-7
NOx	1,800	1,070	107	NA
PM	18,300	3	0.3	~3
PM10	13,700	3	0.3	NA
PM2.5	9,200	3	0.3	NA
Pb	104	0.058	0.0058	0.013~0.046
Hg	2.8	0.018	0.0018	0.1-0.2
Cd	3.4	0.005	0.0005	0.001-0.004
Arsenic	2.14	0.006	0.0006	0.002-0.007
Dioxin	3.5 mg	52.5 ng	5.25 g	NA

Cd = cadmium; dioxin = polychlorinated dibenzo-dioxins and polychlorinated dibenzo-furans; EU = European Union; g/GJ = gram per gigajoule; g/t = gram per ton; Hg = mercury; mg = milligram; NA = not applicable; ng = nanogram; NOx = nitrogen dioxides; Pb = lead; PM10 and PM2.5 = particulate matters with diameter of 10 and 2.5 micrometers or less respectively; PRC = People's Republic of China; SO_2 = sulfur dioxide; WtE = waste to energy.

[a] European Environment Agency guidebook 2023. US EPA (1996).

[b] Converted based on the heat value of 10.5 GJ/t of MSW in the EU countries, same as the US EPA.

Estimation by Analogy

Many incineration plants use actual emission data of existing similar plants to predict their likely emissions (i.e., analogous approach). This is valid especially when there are many incineration plants in operation, making such information more readily available, e.g., in the PRC. The plants did so in their EIA, summarized in Table 11.

Table 11: Actual Incineration Emissions at Discharge (mg/m^3)

Pollutant	Zhoukou 2020[a]		Emission Standard	
	Before Treatment	After Treatment	in PRC[a]	In EU[a] 2010
PM	1,690-1,860	1.8-2.6	<20; 30	<10; 30
SO_2	193-1,076	5-18	<80; 100	<50; 200
NOx	115-138	115-120	<250; 300	<200; 400
CO	10-200	1.26-5.44	<80; 100	<50; 100
HCl	386-393	5.6-7.6	<50; 60	<10; 60
Hg	0.004-0.0044	0.001	<0.05	<0.05
Pb	3.27-3.46	0.1	<1	NA
Heavy metal	0.16-0.17	0.037	<0.1	<0.5
Dioxin[b]	NA	0.0024	<0.1	<0.1

Cd = cadmium, EU = European Union, HCl = hydrogen chloride, Hg = mercury, mg/m^3 =milligrams per cubic meter , NA = not applicable, NOx = nitrogen dioxides, PM = particulate matter, Pb = lead, PRC = People's Republic of China, SO_2 = sulfur dioxide.

[a] For emissions standard for incineration in the PRC (GB18485-2014) and the EU (2010): smaller value is the limit for the 24-hour average, and bigger ones for the 1-hour average. For heavy metals in the EU, 0.5-8 hours.

[b] Dioxin limits are all in International toxic equivalent quantity of nanograms per cubic meter (I-TEQ ng/m^3).

Source: PRC (2014), EU (2020), Zhoukou (2020), which cited data from other Chinese incineration plants.

From the experience of plants in the PRC (Weishi 2020, Zhoukou 2020), incineration usually generates particulate matter about 2%–4% of MSW by weight. Treating 1,200 t/d thus emits 24–48 t/d. Divided by the flue gas flow rate obtained from the feasibility study, the concentration of particulate matter before treatment is estimated. The removal efficiency of the dedust system with a filter bag is usually more than 99%. Thus, particulate matter emission after treatment can be calculated and compared with the applicable emission standard (last column in Table 11).

Estimation by Mass Balance Formula

As incineration with WtE is essentially a thermal power plant using MSW as fuel, it can use the same formula and methods of thermal power plant to estimate quantitatively most pollutants of flue gas emissions (details in Chapter 6). For example, EIAs for incineration plants in the PRC used the same formula of a thermal power plant to quantify particulate matter (PM) in flue gas emissions.

The EIA for Maldives WtE (ADB 2020) included a mass balance provided by the feasibility study. It showed that for the incineration of 500 t MSW/d, flue gas emissions contained 0.14 t/d (or 5800g/h) of sulfur and carbon monoxide (CO) each. Divided by the flow rate (115,712 m^3/h) given by the feasibility study, the carbon monoxide concentration at the stack was obtained (50 mg/m^3, which is the same as the EU emission standard's value). The same mass balance also shows the quantity of fly ash and bottom ash (Table 12).

Table 12: Share of Major Solid Waste from Incineration

Main Waste	Weishi County 2020	Zhoukou 2020	ADB 2019	ADB 2020
Capacity (t/d)	1,200	1,200	400	500
Output (MW)	25	25	8	10
Bottom ash and slag (% of MSW)	23	18–25	20	20
Fly ash (% MSW)	3	~4	3	3
Total (% MSW)	25	22–29	23	23

MSW = municipal solid waste, MW = megawatt, t/d = tons per day.

Odors and Other Air Pollutants

Odor and particulate matter come from non-point sources in incineration plants, such as a garbage pit (bunker) at pretreatment. To control odor, negative pressure is maintained in the waste pit and leachate tank, with the differential pressure always maintained with the adjacent zone to minimize leakage.

Solid Waste from Incineration

The Weishi incineration plant of 1,200t MSW/d will generate slag about 280 t/d (23% of MSW weight) and 33 t/d fly ash (3% of MSW) according to its EIA. Therefore, the total residue that needs to be disposed of is about 310 t/d, or 26% of the MSW wet weight incinerated. The EIA of the incineration project in Viet Nam (ADB 2019) and Maldives

(ADB 2020) had roughly the same predictions for slag (20% of MSW treated) and fly ash (3% of MSW), both based on data from existing similar projects and actual operation experience. Similar percentages and ratios between fly ash and bottom ash can be found in other incineration plants studied (Table 12).

With total ash totaling 25% of MSW by weight still needing disposal, incineration can only reduce MSW by 75%, lower than the reduction achieved by incineration in developed countries, again due to higher moisture and nonflammables of MSW in most developing countries. Other smaller waste streams include leachate treatment sludge and various hazardous waste. According to the EIA for Zhoukou incineration (Zhoukou 2020), leachate treatment sludge is estimated as 4.5 t/d (80% water content) out of 300 m3/d leachate. This unit sludge amount is much higher than that of sewage or general wastewater treatment (Chapter 2) due to the much higher pollution concentration of MSW leachate.

CHAPTER 6

Quantifying Pollution from the Thermal Energy Sector

Thermal energy production mainly takes place in a boiler, furnace, or turbine, and all of which is controlled combustion process. A combination of these combustion devices will form thermal power plants, cogeneration of heat and power, and incineration using different fuels, including waste. They also constitute common types of energy or urban projects of IFIs. Noted for their air emissions, thermal energy production is among the major contributors to air pollution and thus targeted by regulators worldwide. They also generate large amounts of wastewater and solid waste.

Quantifying Flue Gas Emissions

The air emissions of combustion refer to point source pollution discharged from the stack, more commonly termed flue gas emissions. In comparison, fugitive emissions are negligible for combustion plants according to the EEA. In the past decade or so, IFIs—including ADB—have moved away from coal-based energy projects. As a result, this chapter focuses on gas and biofuel-based combustion. For solid fuels such as biomass or MSW, the following formulas, originally developed for coal, can still be used to estimate air emissions and solid waste, but the values for some variables differ.

Estimation by Mass Balance Formulas

Sulfur Oxides Emissions

The emissions of sulfur oxides (SOx) are directly related to the sulfur content of fuel, which for coal normally varies 0.3 -1.2 wt%, fuel oil 0.3%-3%, and negligible for refined natural gas. Most SOx is sulfur dioxide (SO_2) with small proportions of sulfur trioxide (SO_3). Given SO_2 predominantly comes from the oxidation of sulfur in fuel, its emission can be estimated by the chemical composition of fuels (called stoichiometric) based on the mass conservation principle (MEE 2018):

$$M_{SO2} = 2 \times B \times S_{t.ar} \times (1-q_4) \times K \times (1-\eta) \qquad (6\text{-}1)$$

where M_{SO2} is discharge of SO_2, ton/h; *B* is fuel consumption under continuous max output, t/h; q_4 is heat loss due to incomplete combustion, % (Table 13); $S_{t.ar}$ is total sulfur content, % (Table 13); *K* is % of fuel S oxidized into SO_2, % (provided by manufacturer, most around 80%); and η is flue gas desulfurization (FGD) efficiency, % (Table 14).

The formulas also capture key factors affecting flue gas emissions, mainly combustion conditions, which in turn are affected by types of fuel, technologies, and operating practices. The most commonly used furnaces in developing countries include stoker or grate firing technologies, whereas fluidized bed combustion is increasingly used in bigger countries and cities.

Table 13: Reference Value of Key Parameters for Major Fuels

Boiler Type	Coal	q_4 (%)	Sulfur (%)	Ash (%)	Heat Value (GJ/t)
Dry bottom stoker pulverized coal	Anthracite	4.0	About 0.6	6–20	32–33
	Bituminous	1.5–2.0	1–3	Up to 12	18–23
	Lignite	0.5	0.5–3	10–25	17–18
	Washed coal	2.5–3.0	Reduce by 1/3	10–18	NA
Fluidized bed boiler	Anthracite	2.5–3.5	NA	NA	NA
	Bituminous	2.0–2.5	NA	NA	NA
Other fuels					
Oil boiler	Fuel oil, diesel	0	0.5–3.5	0	39–43, 45 (diesel)
Gas boiler	Natural gas	0	Minor	0	38–55 MJ/m³
Solid fuel boiler	Biomass	NA	<0.2	Wood ~1	NA
	Agrowaste	NA	NA	20±	NA
Incinerator with energy recovery	MSW	NA	0.14–0.7 Average at 0.4	11–20	10 in Europe 4–6 in developing countries

GJ/t = gigajoules per ton, MJ/m³ = megajoules per cubic meter, MSW = municipal solid waste, NA = not applicable or available, q_4 = heat loss due to incomplete combustion (in %).

Source: Literature search, except q_4 values which are from MEE (2018).

Table 14: Efficiency of Major Flue Gas Emission Treatment

Desulfurization			Particulate Matter Removal		
Major Technology	Removal (%)[a]	Removal (%)[b]	Major Technology	Removal (%)[a]	Removal (%)[b]
Wet scrubbing by limestone	Up to 98	95.0–99.7	Dry (ESP)	96.5–99.95	99.2–99.9
Ammonium method	NA	95.0–99.7	Wet scrubber/ESP	98.5–99.9	70.0–90.0
Circulate fluidized bed	NA	93.0–98.0	Bag filter	99.6–99.95	99.5–99.90
Seawater method	Up to 90	95.0–99.0	ESP and bag filter combined	NA	99.5–99.99

NA = not available or applicable.

[a] International Finance Corporation. 2007. EHS guidelines for thermal power.

[b] MEE. 2018. Technical guidance on estimation of pollution intensity at source—thermal power, HJ888-2018R. It indicates that wet desulfurization can also remove 50%–70% of particulate matter concurrently.

Particulate Matter

The particulate matter (PM) parameters normally include total suspended particulate (TSP), PM10, and PM2.5, which can represent the total primary PM emission or the filterable PM fraction. Black carbon is also monitored in many countries, as it is an indicator of air pollution and GHG emissions. Since PM control technologies will also reduce black carbon emissions, in general, it is assumed that black carbon can be reduced proportionally to PM2.5 emissions. Consequently, for the EEA air pollution inventory, the black carbon emission factors are expressed as a percentage of PM2.5.

Since the ash content of fuel remains largely inert (not oxidized during combustion), it will either be carried away by flue gas as particulate matter or captured by a dust removal system as fly ash and precipitated as bottom ash and slag. Formula (6-2) is for PM in flue gas (MEE 2018):

$$M_{pm} = B \times Aar \times afh \times (1- \eta d) \quad (6\text{-}2)$$

where M_{pm} is discharge of particulate matter, t/h; *Aar* is ash content of fuel as received, %; *afh* is ash carried by flue gas, % (about 80%-85% or provided by boiler manufacturer); and ηd is removal efficiency of dedusting system, % (Table 14).

Fluidized bed combustion operates at lower temperatures (700°C-900°C) than stoker-type pulverized coal boilers (above 1,300°C), utilizing a bed of sand or other material that suspends fuel particles. Its lower temperature allows in situ desulfurization, which injects sulfur-capturing sorbent directly into the combustion chamber. Limestone or lime is the common sorbent, resulting in more ash than estimated stoichiometrically. However, the overall impact on fly ash is not the primary concern, as it only increases the amount of fly ash in flue gas by typically a few percent, depending on the specific method and conditions.

Heavy Metals

Most of the heavy metals considered are normally released as compounds (e.g., oxides, chlorides) in association with particulates. Only mercury and arsenic are at least partly present in the vapor phase. The content of heavy metals in coal is normally several orders of magnitude higher than in oil products. For natural gas, emissions of mercury and arsenic are relevant. During combustion, particles undergo complex changes that lead to the vaporization of volatile elements. The rate of volatilization of heavy metal compounds depends on fuel and technology characteristics, e.g., boiler type, operating mode, and abatement technology.

The method for estimating is stoichiometric (formula [6-3]), as heavy metals in flue gas emissions are predominantly determined by their content in fuel, as are SO_2 and particulate matter. Their content in fuel received has been routinely analyzed in the laboratory and usually provided by fuel suppliers as well. To control air pollution from combustion, many countries have quality standards for major fuels that specify the maximum percentage of main heavy metals allowed.

$$M_{hm} = B \times m_{hmx} (1- \eta hm) \quad (6\text{-}3)$$

where M_{hm} is the emission of a heavy metal and its compounds (e.g., mercury or lead), g/h; m_{hm} is the content of a heavy metal in fuel as received, mg/kg (or g/t); and ηhm is the joint removal rate of heavy metal by flue gas treatment system, %.

The top concern is mercury because of its toxicity and widespread presence in major types of fuel and waste. Experience shows that desulfurization, denitrification, and dedusting systems can also significantly remove mercury and its compounds, up to 70% on average.

Estimation by Analogy and Emission Factors

Some countries (e.g., the US) and regions (e.g., Europe) have developed emission factors for different types of fuels using main types of boilers and technologies since the 1980s and have kept updating to reflect technological advances. When employing this method, however, emission factors represent long-term averages for all facilities in a particular source or subsector. Thus, it is always better to use those derived from or as close as possible to local conditions, and if unavailable, use emission factors found in literature but supplemented by a survey during project preparation.

As technology characteristics and combustion conditions vary by country and plant, the EEA emission factors represent a mean for the range of technologies in use, with 95% figures in the sector. They do not represent the best available technology nor unabated emissions. The emission factors scattered across several EEA guidebooks are put together in Table 15. One can also consult the AP-42 series on air emission factors for external combustion sources (US EPA).

Table 15: Air Emission Factors for Combustion by Major Fuels

Technology	All Types[a]						Fluidized Bed		Turbine		Incineration[b]	
Parameter	Coal, Lignite	Hard Coal	Heavy Fuel Oil	Natural Gas	Solid Biomass	Biogas	Hard Coal	Lignite	Fuel Gas	Gas Oil	Modern with WtE	No WtE, Old
NOx (g/GJ)	247	209	142	89	81	198	82.5	60	48	398	107	180
CO (g/GJ)	8.7	8.7	15.1	39	90	156	313	13	4.8	1.49	4	70
NMVOC (g/GJ)	1.4	1.0	2.3	2.6	7.3	10	0.9	1	1.6	0.19	0.6	2
SOx (g/GJ)	1,680	820	495	0.24	10.8	10.8	820	1,680	0.24	46.5	8.7	170
TSP (g/GJ)	11.7	11.4	35.4	<0.14	172	NA	8.4	10.2	<0.2	1.95	0.3	1,830
PM10 (g/GJ)	7.9	7.7	25.2	<0.14	155	NA	7.7	6.9	<0.2	1.95	0.3	1,370
PM2.5 (g/GJ)	3.4	3.4	19.3	<0.14	133	NA	5.2	2.8	<0.2	1.95	0.3	920
BC (g/GJ)	2.2[c]	2.2[c]	5.6	NA	3.3[c]	NA	2.2	NA	NA	33.5[c]	3.5[c]	3.5[c]
As (mg/GJ)	14.3	7.1	3.98	<0.12	9.46	<0.042	NA	NA	NA	NA	0.6	214
Pb (mg/GJ)	15	7.3	4.56	<0.0015	20.6	<0.005	NA	NA	NA	NA	5.8	10,400
Cd (mg/GJ)	1.8	0.9	1.2	<0.0003	1.76	<0.002	NA	NA	NA	NA	0.5	340
Hg (mg/GJ)	2.9	1.4	0.34	<0.05	1.51	<0.12	NA	NA	NA	NA	1.8	280
PCDD/F (ng I-TEQ/GJ)	10	NA	2.5	NA	50	<0.96	NA	NA	NA	0.5	5.25	350

As= arsenic; BC = black carbon; Cd=cadmium; CO = carbon monoxide; g = gram; GJ = gigajoule; Hg= mercury; I-TEQ = international toxic equivalent quantity; mg = milligram; ng = nanogram; NA = not available or applicable; NOx = nitrogen oxides; NMVOC = non-methane volatile organic compound; Pb= lead; PCDD/F= polychlorinated dibenzo-dioxins and polychlorinated dibenzo-furans; PM10 or PM2.5= particulate matter with a diameter of 10 or 2.5 micrometers or less, respectively; SOx = sulfur oxides; TSP= total suspended particulate; WtE = waste-to-energy.

[a] The European Environment Agency (EEA) guidebook does not clarify the type of boilers for most fuels, though the commonly used type is dry bottom.

[b] The originals in the guidebook are in kilograms per ton of municipal solid waste. They are converted by the author to g/GJ for easier comparison with other fuel types.

[c] These emission factors for clack carbon are in percentage of PM2.5.

Source: EEA. 2023. *EMEP/EEA Air Pollutant Emission Inventory Guidebook 2023*.

Particulate Matter from Liquid and Gaseous Fuel Combustion

Liquid and gaseous fuels like oil and natural gas hardly contain any ash content. The above formula thus cannot be applied. For oil, the main sources of PM include soot from incomplete combustion, heavy metals, and sulfur in oil (forming sulfate), other impurities, and hydrocarbons combined to form secondary inorganic aerosols in the air, all precursors of PM. For gas-fired plants, the most significant source of PM is the formation of secondary PM. Pollutants emitted, such as nitrogen oxides and volatile organic compounds, react with other chemicals in the air to form fine particles. The processes are highly complex with a lot of uncertainty that renders their estimation by formula impossible. Therefore, the analogy method or emission factors (see below nitrogen oxides emissions and Table 15) have been widely used and accepted.

For solid fuels, including biomass, agrowaste, or MSW, PM emissions come predominantly from the ash content in the fuel. Although the above PM formation processes also exist, their amount can be negligible in practice compared to PM from the ash content of fuel. Therefore, formula (6-3) can apply to these fuels to estimate PM in their combustion emissions.

Nitrogen Oxides Emissions

Nitric oxide (NO) together with NO2 (normally expressed as nitrogen oxides, NOx), under the temperature range of combustion (up to 1,700°C), are formed by the conversion of chemically bound nitrogen in fuel and the fixation of atmospheric nitrogen. Most NOx from coal or solid fuel combustion (80%-90%+) is from fuel nitrogen. The content of nitrogen in solid fuels varies from 0.2-3.5 wt% for coal, 0.1-0.3 wt% for wood, and 0.3-1.4 wt% for waste, much lower for liquid fuel (heavy oil 0.1-0.8 wt%, and fuel oil 0.005-0.07 wt%). Natural gas contains no bound nitrogen, thus only thermal-NOx emissions result from its combustion (EEA 2023 guidebook on energy).

Given that the sources of NOx are not just fuel itself and the share varies drastically across different types of fuel, it is almost impossible to estimate NOx emissions by the stoichiometric method based on mass balance, as in the case of SO_2 and particulate matter. Instead, the analogy and emission factor method has been the common practice in EIAs and is more precise, i.e., using emission concentrations of similar facilities to predict the possible emissions of the one in question.

Quantifying Solid Waste from Combustion

Coal- and biomass-fired thermal power plants generate the most solid waste due to the relatively high percentage of ash in the fuel. The Environmental, Health, and Safety (EHS) Guidelines on Thermal Power Plants (International Finance Corporation [IFC 2007]) provide an example in their footnote 24: a 500 megawatt electric (MWe) plant using coal with 2.5% sulfur, 16% ash, and 30,000 kJ/kg heat content will generate about 500 tons/day solid waste.

There are two types of combustion waste. Large volume waste includes fly ash, bottom ash, boiler slag, and flue gas desulfurization (FGD) sludge. Fly ash removed from exhaust gases makes up 60–85% of the coal ash residue in pulverized-coal boilers and 20% in stoker boilers. Bottom ash includes slag and particles that are coarser and heavier than fly ash. “Low-volume combustion waste” refers to a smaller waste stream consisting of coal mill rejects or pyrites, cooling tower sludge, and sludge from water and wastewater treatment.

Due to much lower sulfur content, oil combustion waste mainly includes fly ash and bottom ash, which are typically substantial when residual fuel oil is burned in an oil-fired steam electric boiler. Other technologies (e.g., combustion turbines and diesel engines) and fuels (e.g., distillate oil) generate little or no solid waste. Overall, oil combustion waste is in much smaller quantities than coal combustion waste.

Gas-fired thermal power plants and boilers generate essentially no solid waste because of their negligible ash and sulfur content, regardless of the combustion technology. Therefore, waste estimation is mainly performed for the coal-combustion process or waste-fired if the waste (including biomass) has a higher ash content.

Fly ash waste estimation uses the same formula for PM in flue gas, except it reverses the removal rate (formula [6-4]). This shows dust removal (like a filter bag) captures particulate matter and turns it into solid waste.

$$M_{fa} = B \times Aar \times a_{fh} \times \eta_d \qquad (6\text{-}4)$$

M_{fa} is fly ash captured by the dust removal system of combustion, t/h.

Bottom ash and slag are essentially the remaining inert component or ash content of fuel after those carried away by flue gases (a_{fh}) and those captured by the dust-removing system (formula [6-5], MEE 2018):

$$M_{bs} = B \times A_{ar} - M_{pm} - M_{fa} = B \times A_{ar} \times (1 - a_{fh}) \qquad (6\text{-}5)$$

where M_{bs} is bottom ash and slag of combustion, t/h.

Flue gas desulfurization sludge is the third component of the large volume waste stream. Wet desulfurization is the predominant method because of its higher desulfurization rate (about 98%) and lower operations and maintenance costs. Limestone or lime slurry is injected to absorb SO_2 in flue gases and form calcium sulfate, also known as gypsum. Formula (6-1) for SO_2 emission applies except that the desulfurization rate is reversed to obtain SO_2 captured by the FGD system instead. The chemical reaction converts SO_2 largely into $CaSO_4$, the main byproduct and bulk of FGD sludge.

$$M_s = 2 \times B \times S_{t.ar} \times (1 - q_4) \times K \times \eta \times C_g \times (1 - C_w) \times 136/64 \qquad (6\text{-}6)$$

M_s is FGD sludge, t/h; C_g is share of gypsum in wet FGD sludge, % (usually about or higher than 90%); and C_w is water content in wet FDG sludge, % (usually less than 10%). The mole weight of $CaSO_4$ is 136, that of SO_2 is 64.

In **dry or semidry flue gas desulfurization**, the chemical reactions result in a mix of products that constitute the FGD sludge, mainly CaSO3 of more than half, gypsum of about 20% with the remainder being limestone formed. The total amount of FGD sludge is usually provided by FGD technology suppliers or estimated by an empirical formula.

In summary, the total amount of solid waste generated by combustion is the sum of formulas (6-4), (6-5), and (6-6). Using the example power plant (500 MW) in the 2007 EHS guidelines, it uses coal with a sulfur content of 2.5%, ash 16%, and a heat value of 30 MJ/kg. Using more conservative values from Tables 13 and 14, fly ash captured by the dust removal system is estimated to be about 7 t/h, accounting for more than half of the solid waste. FGD sludge is 4 t/h by formula (6-6), about one-third of the three sources. Bottom ash and slag are about 1.8 t/h, the smallest.

As power plants typically operate 24 hours a day, the total solid waste from the large-volume stream is the sum of the above three, about 312 t/d [24 x (7 + 4 + 1.8)]. Footnote 24 of the 2007 EHS guidelines state that 500 t/d of solid waste is generated in this example power plant. The remaining 190 t/d comes from the small volume stream.

Estimating Water Pollution

Turbines with boilers and heat recovery generators require a cooling system to condense steam to maximize fuel efficiency. Typical cooling systems used in thermal power plants include (i) once-through cooling systems where sufficient cooling water and receiving surface water are available, (ii) closed circuit wet cooling systems or recirculated cooling, and (iii) closed circuit dry cooling (e.g., air cooled condensers).

The cooling system typically accounts for 95%–98% of the water consumption of a power plant, but with minimal pollutants (EHS guidelines 2007). Thus, in the once-through system, heated water is discharged directly back to the environment, causing thermal pollution of the receiving water body, a major water pollution problem of power plants. Most countries and international standards require the thermal discharge not to exceed 3°C of ambient water temperature.

Recirculated cooling can greatly reduce water consumption and consequently thermal discharge. However, a small fraction of water still needs to be discharged (called blowdown) to prevent the build-up of minerals and other impurities, to prevent scaling, corrosion, and fouling, with makeup water to compensate for loss in blowdown, evaporation, and leaks. The blowdown is typically 5%–10% of cooling water, depending on the water quality and operating conditions.

Process Wastewater

In power plants with recirculated cooling, wastewater streams primarily include cooling system blowdown; water demineralization (or softening) wastewater, boiler blowdown and cleaning, ash-handling wastewater; wet FGD system discharges; and equipment and floor cleaning, laboratory waste, and domestic wastewater. All of these are usually present in plants burning coal, some (e.g., ash-handling wastewater) may be present in reduced quantities, and some (e.g., FGD) may not be present at all in gas- or oil-fired or some biomass power plants. Their amounts vary for different types of fuels, boilers, and operating conditions (Table 16).

The characteristics of the wastewater depend on how the water has been used. Contamination arises from demineralizers; lubricating and auxiliary fuel oils; trace contaminants in the fuel (introduced through the ash-handling wastewater and wet

FGD system discharges); and chlorine, biocides, and other chemicals used to manage the quality of water in cooling systems. Cooling tower blowdown tends to be very high in total dissolved solids, residual chlorine, and toxic chemicals that may be present in cooling tower additives (Table 17).

The approximate scale of wastewater streams provides only an indication due to the big variability and difficulty in deciphering data in the literature, of which the most useful proved to be EIAs. For power plants using coal and oil of higher sulfur content, desulfurization wastewater is usually treated separately from other process wastewater due to the special properties that make it most difficult and costly to treat (Wang et al. 2019).

Table 16: Main Wastewater Streams of Biomass and Gas-Fired Thermal Power Plants

	Biomass Power Plants		Gas-Fired Power Plants		
Wastewater	**World Bank 2010**	**ADB 2018**	**ADB 2012**	**ADB 2017b**	**ADB 2019b**
Designed capacity, fuel composition	30 MW, 2x75 t/h FBB, 990 t/d of wheat straw (ash 8%) and corn stalk (ash 4%), S is 0.07% for both	As left, same in scale, FBB and biomass types	1,700 MW CCGT: 4x285 MW gas boilers +2x300 MW gas turbine	800 MW, natural gas, trace level of hydrogen sulfide	800 MWe, two units, natural gas zero sulfur, high diesel oil for emergency has S<0.25%; air cooling
1. Blowdown of cooling water	1,030 m^3/d	960 m^3/d	7,000 m^3/d	~3,200 m^3/d (in winter)	Unclear but process WWT: 2,640 m^3/d
2. Blowdown of demineralization, boiler system	480 m^3/d	NA	356 m^3/d	~2,700 m^3/d	3,840 m^3/d (160 m^3/h makeup water)
3. Wastewater from flue gas treatment	Negligible, dedust only	Negligible, same as left	NA	Negligible	NA
4. General wastewater	13 m^3/d	12 m^3/d	120 m^3/d	~12 m^3/d	~36 m^3/d

ADB = Asian Development Bank, CCGT = combined cycle gas turbines, FBB = fluidized bed boiler, m^3/d = cubic meters per day, MW = megawatt, NA = not applicable, t/d = tons per day, t/h = steam ton per hour.

Table 17: Properties of Major Wastewater Streams of Thermal Energy Production

Wastewater	Approximate Scale	Typical Property (mg/L)	Typical Treatment
1. Blowdown of cooling water	For biomass and gas: >1,000 m^3/d (Table 16)	**Moderate concentration** Dissolved solid: several hundred	Biochemical (i.e., activated sludge, MBR) followed by ultrafilter and RO
2. Blowdown of water softening, cleansing boiler	Vary; for biomass and gas: several hundreds m^3/d (Table 16)	**Low concentration**[a] BOD: 100–200; SS: 50–100 COD: 100–350; salt: ~3,000 NH_3-N: 10–30	Low concentration WW, often treated in a system separated from leachate: chemical-physical, i.e., flocculation, filtering, RO, or nanofilter
3. Wastewater from flue gas treatment	Negligible to minor for biomass and gas-fired	**High in inorganics**[b] SS: 10,000–15,000 Cl: 5,000–20,000; Ca2+: 1,360 Mg2+: 4,240	
4. Concentrate of membrane tech (RO, ultrafilter)	20%–35% of liquid after RO, various filters	High in salts	Lime slurry making, fly ash stabilizing, back-spray in combustion, evaporation
5.Domestic/general: cleaning vehicle /area, staff use	Small	Chapter 1 and the 2023 ADB study of this series	Drained to sewer, septic tank, or leachate system

ADB = Asian Development Bank, BOD = biological oxygen demand, Ca2+ = Calcium (II) ion, Cl =chloride, COD = chemical oxygen demand, m^3/d = cubic meters per day, MBR = membrane bioreactor, Mg2+ = manganese (II) ion, mg/L = milligrams per liter, NH3-N = ammonia-nitrogen, RO = reverse osmosis.

[a] ADB (2023).

[b] Wang et al. (2019).

CHAPTER 7

Conclusion

Most projects generate pollution, especially during their long-term operation. Almost all countries have discharge standards for emissions and effluents that are quantitative and concentration based. This is one of the main reasons many countries require EIAs to predict and estimate pollution quantitatively to demonstrate whether pollution from a project can comply with the applicable standard or not. If pollution is assessed to fail to meet the standard, control methods and treatment need to be identified and justified, again entailing quantification, if they can bring pollution into compliance.

With methods to quantify and measure point source pollution becoming more mature and widely accepted over the years, quantitatively estimating it has become a standard requirement for EIAs, even in many developing countries. Discharge standards are also developed mainly for point source pollution, as one can only manage (and enforce laws for) what can be measured.

Semi- and non-point source pollution is more complex and therefore much harder to quantify and measure, as shown in the chapters on waste landfill and biotreatment. Nonetheless, efforts have been made to quantify these sources of pollution to the extent possible, especially for those that have discharge standards.

An intriguing phenomenon was uncovered during the research of EIAs for this study. For flue gas emission prediction, some EIAs state "Modern manufacturers guarantee (or Design–Build–Operate [DBO] contract mandates) that NOx emissions be less than 50 mg/m^3 (the standard adopted by the project) given its design and technologies ... " Other EIAs state "NOx emission is 50 mg/m^3 at stack based on calculation by xx" without any explanation on how this number is arrived at and justified. These emission standards values are then used as input to dispersion models (usually AERMOD, the US EPA's air quality model) to predict ground level concentrations at various distances from the source and to calculate the stack height.

This essentially involves modeling the emission standard, rather than the predicted emissions themselves. It often occurs in gas-fired power plants, where the primary air emission is NOx. A survey in the PRC (North Star Power Net 2018) revealed that most gas power plants have consistently met the emission standards even without desulfurization or denitrification (Table 18). Emission standards of the PRC are the same as those of the EU for gas-fired power plants. Given the situation, modeling the value of emissions standards in EIAs has been permitted in recent years, unlike in the past. The authors of the 2018 survey went on to advocate tightening the emissions standards for gas-fired power plants.

In some cases, emissions data from similar plants nearby are readily accessible, or from earlier phases or trial operations. Such data can well be used as analogy to infer a project's emissions. Yet their EIAs did not do so nor explain why. For those that did explain, the main reason given was, "to be more conservative in case of the worst scenario." The example in Box 2 shows that the input chosen for modeling was two to eight times the actual emissions, well beyond what is typical for sensitivity analysis of abnormal conditions.

Table 18: Air Emission of Gas-Fired Power Plants in the People's Republic of China (mg/m3)

Pollutant	Beijing	Yangzi River Delta	Pearl River Delta	Standard for Gas-Fired Power
Nitrogen oxides	<25	<50	<50	50
Sulfur dioxide	<5	<5	<5	35
Particulate matter	<3	<3	<3	5–10
Denitrification	yes	Only two plants	none	none

Note: None of the surveyed units has a desulfurization or dedusting system.

Source: North Star Power Net (2018).

Box 2: Modeling Emissions Versus Modeling the Emissions Standard

A private sector incineration project in Viet Nam has a capacity of 400 t/d MSW (ADB 2019a). The owner started trial operation in late 2018, with continuous flue gas monitoring data provided in the EIA. The project adopted the EU standard (EU 2010) because it is more stringent than the domestic one, both of which use limits on a 24-hour average. However, for input to the dispersion modeling, domestic standards values were used, plus a 20% margin (see table below), to be more conservative for the worst-case scenario.

Pollutant	Actual Emissions (mg/m^3)	EU Standard (mg/m^3)	Domestic Standard (mg/m^3)	Input to Model (mg/m^3)
SO_2	15–45	50	250	300
NOx	170–390	200	500	600
TSP	All <10 except one sample (18)	PM <10 None on TSP	120	150
HCl	10–45	10	50	60

In sharp contrast to the overestimation above, other projects underestimated future emissions and thus impacts on air quality. Despite emissions from similar projects higher than applicable emission standards according to their EIAs, they still used the values in the emission standard as input to dispersion modeling. There are usually no explanations or justifications, just a statement like, "the contractor is bound by Design-Build-Operate contract to guarantee compliance with such standard," or "Modern gas turbine manufacturers guarantee the formation of NOx less than ...(the value in emission standard) due to their improved firing system etc.," or "based on xx firm's calculation."

ADB = Asian Development Bank, EIA = environmental impact assessment, EU = European Union, HCl =hydrogen chloride, mg/m3 = milligrams per cubic meter, MSW = municipal solid waste.

NOx = nitrogen oxides, PM = particulate matter, SO_2 = sulfur dioxide, TSP = total suspended particulate.

Source: Author.

The harm of underestimating being needless to say, this practice even when overestimating can lead to oversized stack height and design, inflating the investment cost unnecessarily, and may also undermine the judgment of environmental agencies and relevant law enforcement. For example, if the EIAs and its emission modeling could use actual data from existing operations to infer the project's emissions rather than relying on the emission standards' values, government agencies could see the need to tighten the emission standards more quickly.

In India, the Central Pollution Control Board (CPCB) issued a series of standards and norms on thermal power plant emissions and effluent. For plants with wet flue gas desulfurization, the notification provided simple empirical formulas for estimating stack height and minimum height (CPCB 2018), as well as benchmarks for water consumption of the main processes. These can greatly assist technical designers and EIA preparers facing time and resource constraints, particularly in developing countries, and can also help crosscheck the modeling results.

In most Design–Build–Operate or similar contracts, using the applicable discharge or quality standard may be the only approach feasible because that is all that can be contractually binding on technical design. However, any targets in the contracts for detailed design and execution are after thus based on the findings and results of feasibility study and on environment-related matters, also EIA. Therefore, the EIA and feasibility study preparers need to improve quantification of emissions at source or effluents at outfall, and document the analysis process and results better in their reports. Modelers just get the estimated emissions, effluents, or other instructions and assumptions from the feasibility study or EIA as input to run the computerized models.

Emissions are greatly affected by operational conditions, even when state-of-the-art technology and equipment are used. The guaranteed performance, even by contractors with good track records, entails operating conditions that are not always obtainable or stable in developing countries with limited experience. Thus, a sounder way and better practice for estimating and predicting emissions remains those described in Chapters 1–6.

While this study focuses on the quantitative aspects of environmental work, technical studies should not be carried away single-mindedly by the pursuit of more precise quantification. Often, a rough estimate based on a rule of thumb or empirical formula is sufficient to make judgments on future compliance and the likelihood of certain treatments to bring pollution into compliance. The more important thing is not to lose sight of the bigger picture, but to ensure that the pollution controls and treatments recommended by the EIA, with its analysis and quantification, can be realized in the project. Otherwise, EIA efforts will be wasted, regardless of how accurate the quantification is or how advanced the techniques or models are.

EIA work is an iterative process that needs to feed into the technical design of the feasibility study in a timely way. Some feasibility study experts accept whatever the project proponent proposes as the basis for design without conducting their own analysis, regardless of whether such a request makes sense. Accepting and copying what is in a feasibility study by EIA preparers without analysis will repeat the mistakes and fall short of fulfilling the role of an EIA.

Considering the end goals of the EIA process (Chapter 1), it is not EIA's task to assess every impact or risk, nor is it practical. This can overwhelm environmental professionals with little time and energy left to ensure whatever is in the EIA can really be heeded and adopted by the feasibility study, thus increasing the chance of being executed. The EIA is not just a step needed to get project approval, and then largely shelved afterward.

The roles of EIAs are to aid decision-making and inform design to avoid and minimize negative project impacts (mitigating impacts is secondary). One can only play such roles at the feasibility study stage before the high-level decision and design are made. This is why most countries require EIAs to be conducted during the feasibility study stage, and many explicitly say so in their regulations. Otherwise, it will be too late and too costly to change course or improve the design.

According to the management hierarchy, it is more cost effective to avoid and minimize negative impacts like pollution first and foremost through high-level decisions or design at the feasibility study stage, such as changing the scale, adjusting the alignment, or even not doing a project. For residual impacts that cannot be avoided or minimized through design optimization, mitigate them with measures such as pollution treatment. If the primary role of the EIA process is merely to mitigate impacts throughout a project cycle, this is essentially downgrading EIA to greenwashing, which can fail to achieve even mitigation itself. As the famous military strategist of ancient China, Sun Tzu, wrote more than 2,000 years ago in his classic, *The Art of War*, "Aim for the high, and you may achieve the middle; aim for the middle, you may achieve the low."

APPENDIX

Soil Quality Standard and Discharge Standard

Many countries have soil quality standards. The report by researchers from the Netherlands is a good source as it compiles the standards in some developed countries in Europe and North America (Tamis and Karman 2008). Table A1 offers an example from developing countries. The numeric values in standards of other countries are similar, though classified differently, e.g., by acidic, neutral, and alkalic conditions of soil in the case of the People's Republic of China (PRC).

Not all countries have standards for land-based disposal of sludge from wastewater or other pollution treatment and sediment dredged, even less for the utilization of these materials. Tables A2 and A3 provide some examples.

Table A1: Soil Standards in Viet Nam, 2015

	Maximum Limit by Land Type (all in mg/kg of dry soil)				
Parameter	Agricultural	Forest Land	Inhabited	Industrial	Trade Service
Cadmium (Cd)	1.5	3	2	10	5
Mercury (Hg)	NA	NA	NA	NA	NA
Lead (Pb)	70	100	70	300	200
Chromium (Cr)	150	200	200	250	250
Arsenic (As)	15	20	15	25	20
Copper (Cu)	100	150	100	300	200
Zinc (Zn)	200	200	200	300	300
Nickle (Ni)	NA	NA	NA	NA	NA

Source: ADB. 2024. *Climate Adaptation through Irrigation Modernization Project: Updated Initial Environmental Examination.*

Table A2: United States Standards on Land Application of Sludge (Excerpt) (all max. in dry weight)

Pollutant	Ceiling Limit (mg/kg of soil)	Monthly Average (mg/kg of soil)	Cumulative Load (kg/ha)	Annual Load (kg/ha/365 days)
Cadmium	85	39	39	1.9
Mercury	57	17	17	0.85
Lead	840	300	300	15
Chromium	3,000	1,200	3,000	150
Arsenic	75	41	41	2.0
Copper	4,300	1,500	1,500	75
Zinc	7,500	2,800	2,800	140
Nickle	420	420	420	21
Selenium	100	36	100	5

kg/ha = kilograms per hectare, mg/kg = milligrams per kilogram.

Source: United States Environmental Protection Agency. 1994. Land Application of Sewage Sludge, EPA/831-B-93-002b.

Table A3: Standards for Land Application and Utilization of Sludge in the People's Republic of China (all max. in dry sludge or sediment, mg/kg)

Pollutant	For Agriculture Use[a]		In Forest Land[b]	For Brick Making[c]	In Cement Production[d]
	pH <6.5	pH ≥6.5			
Total cadmium	5	20	20	20	20
Total mercury	5	15	25	5	15
Total lead	300	1,000	1,000	300	1,000
Total chromium	600	1,000	1,000	1,000	1,000
Total arsenic	75	75	75	75	75
Total copper	250	500	1,500	1,500	1,500
Total zinc	500	1,000	4,000	4,000	3,000
Total nickel	100	200	200	200	200
Mineral oil	3,000	3,000	NA	3,000	3,000
Benzopyrene	3	3	NA	NA	3
PAHs	NA	NA	NA	40	6

NA = not available, PAH = polycyclic aromatic hydrocarbon.

[a] Control standard for pollutants in sludge for agricultural use (GB4284-84), 1984.

[b] Quality of sludge used in forest land (CJ/T362-2011), 2011.

[c] Quality of sludge used in brick making (GBT25031-2010), 2010.

[d] Quality of sludge used in co-production of cement clinker (CJT314-2009), 2009.

Source: Unofficial English translation by the author of the above standards from the People's Republic of China.

REFERENCES

Asian Development Bank (ADB). 2012. *Final Environmental Impact Assessment: Ayudhaya Natural Gas Power Project in Thailand.*

ADB. 2017a. *Environmental Impact Assessment (Main Report): Power Generation Efficiency Improvement Project in Uzbekistan*.

ADB. 2017b. *Guidelines for Estimating Greenhouse Gas Emissions of Asian Development Bank Projects: Additional Guidance for Clean Energy Projects*.

ADB. 2018. *Environmental Impact Assessment: Air Quality Improvement in the Greater Beijing–Tianjin–Hebei Region — Shandong Clean Heating and Cooling Project (Biomass Combined Heating and Power Generation Component)*.

ADB. 2019a. *Environmental and Social Impact Assessment: Municipal Waste-to-Energy Project in Viet Nam*.

ADB. 2019b. *Updated Environmental Impact Assessment: Rupsha 800 MW Combined Cycle Power Plant Project in Bangladesh*.

ADB. 2020. *Environmental Impact Assessment: Greater Malé Waste-to-Energy Project in the Republic of Maldives — Waste to Energy Plant PART A*.

ADB. 2023. *Pollution Control Technologies for Small-Scale Operstion*.

ADB. 2024. *Pollution Characterization and Quantification in the Agriculture Sector*.

Animonan Energy Inc. 2023. *Project Description for Scoping A1E 4x600MWnet Natural Gas-fired CCGT Power Plant Project, Philippines*.

Beijing 2008. *Environmental Impact Assessment of the Integrated Municipal Solid Waste Treatment Facility in Baigou Town of Gaobeidian City*.

Cai, B.-F. et al. 2014. *Estimation of Methane Emissions from Municipal Solid Waste Landfills in China Based on Point Emission Sources*. *Advances in Climate Change Research*. 5 (2). pp. 81–91.

Central Pollution Control Board (CPCB), India. 2018. *Thermal Power Plant Effluents and Emissions Standards*.

Chen, Z. 2022. *"Zero Discharge" Technology and Engineering Case Analysis of Wastewater Treatment of Solid Waste Incineration Power Plants* (in Chinese). *Water Purification Technology*. 41 (7). pp. 99–105.

Cheng, S.et al. 2023. *Immigration, Transformation, and Emission Control of Sulfur and Nitrogen During Gasification of MSW: Fundamental and Engineering Review*. *Carbon Resources Conversion*. 6 (3). pp.184–204.

Choden, Y. et al. 2022. *Modeling of Leachate Generation from Landfill Sites*. *Nature Environment and Pollution Technology*. 21 (3). pp. 993–1002.

Environment Australia. 1999. *Emission Estimation Technique Manual for Sewage Sludge and Biomedical Waste Incineration*.

Environment Canada. 2013. *Technical Document on Municipal Solid Waste Organics Processing*.

Environment Network. 2023. *Current Status and Case Analysis of MSW Composting in Rural Areas* (in Chinese).

European Environment Agency (EEA). 2019. *European Monitoring and Evaluation Programme (EMEP)/EEA Air Pollutant Emission Inventory Guidebook 2019*.

EEA. 2023. *EMEP/EEA Air Pollutant Emission Inventory Guidebook 2023*.

European Union. 2010. *Directive 2010/75/EU of the European Parliament and of the Council of 24 November 2010 on Industrial Emissions (Integrated Pollution Prevention and Control)*. In *Annex VI on Technical Provisions Relating to Waste Incineration Plants and Waste Co-Incineration Plants*.

Intergovernmental Panel on Climate Change (IPCC). n.d. *Emissions from Waste Incineration*. In *Good Practice Guidance and Uncertainty Management in National Greenhouse Gas Inventories*. pp. 455–468.

IPCC. 2019. *2019 Refinement to the 2006 IPCC Guidelines for National Greenhouse Gas Inventories*.

International Energy Agency (IEA). 2003. *Municipal Solid Waste and its Role in Sustainability*. A position paper prepared by IEA Bioenergy.

International Finance Corporation. 2007. *EHS Guidelines for Thermal Power Plants*.

International Organization for Standardization (ISO). 2022. *ISO 24297:2022: Guidelines for Treatment and Reuse of Leachate from Municipal Solid Waste (MSW) Incineration Plants*.

Karagozoglu, M. B. and M. Asar. 2023. *Physico-Chemical Characteristics of Leachate from Landfills and Its Impacts on Surface Water Quality: Case Study in Osmaniye/Turkey*. *Sustainable Water Resources Management*. 9. 125.

Lindamulla, L. et al. 2022. *Municipal Solid Waste Landfill Leachate Characteristics and Their Treatment Options in Tropical Countries*. *Current Pollution Reports*. 8. pp. 273–287.

Logan, M. et al. 2021. *Anaerobic Digestion of Dissolved Air Floatation Slurries: Effect of Substrate Concentration and pH*. *Environmental Technology and Innovation. 21. 101352.*

Ministry of Ecology and Environment (MEE). 2008. Pollution Control Standard for Municipal Solid Waste Landfill (GB 2016889-2008). Government of the People's Republic of China (PRC).

MEE. 2013a. Pollution Emission Factors and Calculation Methods (in Chinese). Government of the PRC.

MEE. 2013b. Technical Specification for Sanitary Landfilling of Municipal Solid Wastes (GB 50869-2013). Government of the PRC.

MEE. 2014. *Standard for Pollution Control on the Municipal Solid Waste Incineration* (GB18485-2014). Government of the PRC.

MEE. 2018. Technical Guidance on Estimation of Pollution Intensity at Source—Thermal Power Industry (HJ888-2018) (in Chinese). Government of the PRC.

MEE. 2021. Emission Factors Manual for Centralized Pollution Treatment Facilities. Government of the PRC.

Ministry of Housing and Urban-Rural Development (MOHURD). 2017. *Specifications for Urban Drainage Planning* (GB 50318-2017). Government of the PRC.

North Start Power Net. 2018. *Current Status and Policy Trends in Flue Gas Emissions of Chinese Gas-Fired Power Plants*.

Ren, X. 2021. *Integrated Planning and Policy Design for Sustainable Waste Management*. *Development Asia Insight*. Asian Development Bank. 28 January.

Shi, Weijie et al. 2022. *Optimization and Adjustment of Biogas Slurry from Kitchen Waste* (in Chinese). Advances in Environmental Protection. 12 (3). pp. 554–569.

Solid Waste Net. 2022. *Main Problems in MSW Leachate Treatment and Suggestions* (in Chinese).

Tamis, J. E. and C. C. Karman. 2008. *Soil and Groundwater Quality Standards*.

United States Environmental Protection Agency (US EPA). n.d. *AP 42, Fifth Edition, Volume I Chapter 1: External Combustion Sources*.

US EPA. 1996. Chapter 2.1: Refuse Combustion. In *AP-42: Compilation of Air Emissions Factors from Stationary Sources*. 5th ed. Vol. I.

US EPA. 2016. *Standards of Performance for Municipal Solid Waste Landfills.* Federal Register. 81 (167). pp. 59332– 59384.

US EPA. 2025. *Chapter 2.4: Municipal Solid Waste Landfills.* In AP-42. Vol. I.

Wang, W., M. Xiaodan, and Z. Fanfu. 2019. *Research Progress on Desulfurization Wastewater Treatment Process of Thermal Power Plants.* *Water Pollution and Treatment*. 7 (1). pp. 18–24.

Weishi County. 2020. *EIA of Municipal Solid Waste Incineration to Generate Power in Weishi County, Hebei Province, PRC* (in Chinese).

World Bank. 2010. *EIA of Anqiu Biomass Cogeneration Project*.

World Bank. 2012. *China - Ningbo Municipal Solid Waste Minimization and Recycling Project: Environmental Assessment (Vol. 2 of 2): Environmental Impact Report*.

World Bank. 2014. *Environmental Audit Reports for Kunming XiShan Municipal Solid Waste Incineration Power Plant*.

Xiang, R. et al. 2025. *A Review on Landfill Leachate Treatment Technologies: Comparative Analysis of Methods and Process Innovation*. *Applied Science*. 15 (7). 3878.

Zhang, H. et al. 2011. *Impacts of Agro-Stalks on Odors and Leachate from Kitchen Waste Composting* (in Chinese).

Zhoukou. 2020. EIA of MSW to Energy Project in Zhoukou City Industrial Park (in Chinese).

www.ingramcontent.com/pod-product-compliance
Lightning Source LLC
Chambersburg PA
CBHW041759220926
51847CB00014B/255
* 9 7 8 9 2 9 2 7 7 6 6 5 7 *